Applied Geomorphology for Mitigation of Natural Hazards

Advances in Natural and Technological Hazards Research

VOLUME 15

The titles published in this series are listed at the end of this volume.

Applied Geomorphology for Mitigation of Natural Hazards

by

MASAHIKO OYA

Emeritus Professor of Waseda University,
and Technical Advisor of Nikken Consultants Inc.,
Tokyo, Japan

KLUWER ACADEMIC PUBLISHERS
DORDRECHT / BOSTON / LONDON

A C.I.P. Catalogue record for this book is available from the Library of Congress.

ISBN 0-7923-6719-7

Published by Kluwer Academic Publishers,
P.O. Box 17, 3300 AA Dordrecht, The Netherlands.

Sold and distributed in North, Central and South America
by Kluwer Academic Publishers,
101 Philip Drive, Norwell, MA 02061, U.S.A.

In all other countries, sold and distributed
by Kluwer Academic Publishers,
P.O. Box 322, 3300 AH Dordrecht, The Netherlands.

Printed on acid-free paper

Printed in the Netherlands.

Table of Contents

Foreword ix

1. Geomorphological Survey Maps Showing Classification of Flood-Stricken Areas 1
   1.1. History of Geomorphological Survey Mapping in Japan 1
   1.2. Value of Geomorphological Survey Maps 6
   1.3. Method of Preparing Geomorphological Survey Maps for Alluvial Plains 8

2. Geomorphological Studies in Europe 11
   2.1. Development of Geomorphological Mapping in Europe 11
   2.2. Geomorphological Mapping Working Groups in Europe and Japan 12
   2.3. The International Geomorphological Map of Europe 13

3. Validation of a Geomorphological Survey Map 17
   3.1. Geomorphology of the Nobi Plain 17
      3.1.1. The Ise Bay Typhoon 17
      3.1.2. Features of Flooding 23
   3.2. Application to the Ariake Sea Lowland 25
   3.3. Using Geomorphological Land Classification to Estimate Flooding from Tsunami 28
      3.3.1. Tsunami in Japan 28
      3.3.2. Comparing Tsunami from Near and Afar 29
      3.3.3. Coastal Landforms of Affected Regions 29
      3.3.4. Tsunami Routes 31
      3.3.5. Profile of Inundation 34
      3.3.6. Conclusions 35

4. Relationships Between Geomorphic Units and Flood Types of River Basins 37
   4.1. Geomorphic Features of Japan 37
   4.2. Differences in Geomorphology Caused by Differing Climatic Conditions 39
   4.3. Features of Japan's Alluvial Plains 41
      4.3.1. Basic Form of Alluvial Plains 41
      4.3.2. Influence of Eustatic Movement and Human Activity 42
   4.4. Overflowing Type: The Kiso River 42

4.5. Concentration Type (Retarding Type): The Chikugo River 45
4.6. Combination of Overflowing and Concentration Types: The Kano River 49
4.7. Natural Conditions in the Korean Peninsula 52
4.7.1. Climatic Conditions 52
4.7.2. Geologic Conditions 55
4.8. Overflowing Type: The Nakdong River 55
4.9. Studies of Overflowing-Type Flooding from the Han River 57
4.10. Map Utilization for Deciding Embankments along the Mekong River 61

5. Flooding in Semi-Frigid Zones 67
5.1. The Ishikari River 67
5.2. Rivers and Fluvial Plains along the Okhotsk Sea 69
5.2.1. Features of Group A 71
5.2.2. Features of Group B 72
5.2.3. Common Features of A and B 72
5.3. The Vistula River in Poland 73

6. Differences in Geomorphology and Flooding Between the Left and Right Banks of Rivers along the Median Dislocation Line 79
6.1. Japan's Landforms 79
6.2. The Yoshino River 81
6.2.1. Brief Description of the Yoshino River 81
6.2.2. Historical Records of the Yoshino River 81
6.2.3. Geomorphic Features of the Yoshino River Plain 82
6.2.4. Utilization of Geomorphological Land Classification Maps 86
6.3. The Toyo River 88
6.3.1. Geomorphic Features of the Toyo River 88

7. Estimation of Land Collapse in Japan's Mountainous and Volcanic Regions 91
7.1. Land Collapse in Mountainous Regions 91
7.1.1. Features of Mountains and Volcanoes in Japan 91
7.1.2. Features of the Kiso Mountain Range 94
7.1.3. Land Collapse in the Nakatsu River Basin 94
7.1.4. Relationship Between Geomorphology and Land Collapse 97
7.2. Landslides from Volcanoes 97

7.2.1. Features of Volcanoes in Japan 97
7.2.2. Landslides and Flooding in the Taradake Volcano Area 98

8. Japan's Coastal Lakes 101
8.1. Geomorphic Features of Coastal Lakes 101
8.2. Method of Coastal Lake Map Preparation 103
8.3. Kasumigaura Lake 103
8.3.1. Features of the Kasumigaura Lake Area 103
8.3.2. Topography of the Kasumigaura Lake Bed 105
8.3.3. Geomorphic Development Around Kasumigaura and Kitaura Lakes 106
8.4. Ogawara Lake 106
8.4.1. Features of the Ogawara Lake Area 106
8.4.2. Topography of the Ogawara Lake Bed 108
8.5. Map Utilization 108

9. Flood Control in Tokyo, Osaka and Nagoya 111
9.1. Flooding in the Tokyo Area 111
9.1.1. Outline of Tokyo's Geomorphology 111
9.1.2. Typhoon Catherine 111
9.1.3. The Kano River Typhoon 112
9.1.4. Recent Flooding 112
9.2. Inundation Caused by Rainfall in the Osaka Area 114
9.2.1. Topography of the Osaka Plain 114
9.2.2. Causes of Inundation 116
9.3. Integrated Flood Control: Nagoya 117
9.3.1. Nonstructural Flood Control 117
9.3.2. City Planning for the Mitigation of Flood Damage 117

10. Determining Areas At Risk of Soil Liquefaction During An Earthquake 121
10.1. Geomorphic Features of Sites Affected by Soil Liquefaction 121
10.2. Flooding and Soil Liquefaction in the Shonai Plain 124
10.2.1. Method of Map Preparation 124
10.2.2. Brief Description of the Shonai Plain 124
10.2.3. Relationship Between Soil Liquefaction Caused by the 1964 Earthquake and Geomorphic Units of the Plain 126

10.3. Hyogo Prefecture Nanbu Earthquake of 1995 (The Great Hanshin Earthquake) 126
10.4. Estimating the Potential for Soil Liquefaction from Earthquakes 128

11. Use of Geomorphological Land Classification Maps in Technical Assistance to Developing Countries 131
11.1. Use of a Geomorphological Land Classification Map in the Padang Project 131
11.2. Map Use in Selecting a Site for a Bridge over the Brahmaputra-Jamuna River 135
11.2.1. Methodology 136
11.3. Geomorphological Features of the Brahmaputra-Jamuna-Ganges River Plain 137
11.4. Geomorphic Development of the Brahmaputra-Jamuna-Ganges River Plain 141
11.5. Shifting of the Ganges and Brahmaputra-Jamuna Rivers 142
11.5.1. Shifting of the Ganges River Course 142
11.5.2. Shifting of the Brahmaputra-Jamuna River Course 143
11.6. Selection of the Bridge Site 143

Appendix 147
Appended Table 1 148
Appended Table 2 156
Appended Table 3 157
Appended Figure 158

References 159

Index 163

About the Author 167

Color Plate Section 169

Map (inside cover)

FOREWORD

I have loved geography since I was a child. Furthermore, I wanted to study in a field that would be useful to society and humankind, and I found this in geography.

In 1950, when I was a student of human geography at Nagoya University, I heard a special lecture from Fumio Tada, who was an associate professor at the University of Tokyo. The lecture was on applied geography, with a particular focus on depositional geomorphology of alluvial plains. I was greatly impressed. When I finished my studies at Nagoya University, I entered graduate school at the University of Tokyo and began studying applied geography under Dr. Tada.

In 1956, I created "A Topographical Survey Map of the Kiso River Basin Showing Classification of Flood-Stricken Areas." This type of map not only enables us to determine the features of past floods but also to predict those of future ones. The accuracy of the map was confirmed in 1959, three years after its preparation, when the Ise Bay Typhoon created a tidal wave. The features of the resultant flooding were almost identical to those predicted by the map (see appended map).

Furthermore, I found that the plain consisted of three geomorphic elements: fan, natural levee with back-marsh, and delta. This combination of geomorphic elements is a fundamental feature of the plains in Japan, as well as of depositional plains in other countries.

After that, at the request of the Science and Technology Agency and of the Ministry of Construction, I prepared a map of the important rivers in Japan. Then, at the request of the U.N. Economic Commission of Asia and Far East (ECAFE, now known as ESCAP) and of the Japan International Cooperation Agency (JICA), I prepared a similar map for the important rivers of Southeast Asia (*appended Table 1*).

Dr. Tada, then vice chairman of the International Geographical Union (IGU), organized the IGU Regional Conference of Japan in 1957. He and I delivered a paper on the subject: "Flood Type and Classification of Topography."

I joined the Geographical Survey Institute, Ministry of Construction, in 1959. At that time Japan was poor, and it was very difficult to travel abroad. But fortunately for me, the chances came one after another, and I was able to fulfill my childhood dreams of going abroad and working internationally.

With the help of the U.S. Science Council, I was able to attend the 10th Pacific Science Council in Hawaii in 1961.

In 1960 and 1961, at the request of the Bureau of Flood Control and Water Resources Development, ECAFE, which was chaired by Koichi Aki, I researched deltaic areas in Southeast Asia, including Cho Shui Chi Delta in Taiwan, Chao Phraya Delta in Thailand, Irrawaddy Delta in Burma, and Sundarbans in India and East Pakistan. This project involved three experts from the Netherlands, three from Japan and two U.N. staff members. Leading the group was A. Volker. A soil scientist, I. Zonneveld, and I researched the topography of the plains. This cooperation with foreign experts was a great experience.

In 1964, I attended a symposium on "Humid Tropical Zone Deltas and Their Implications," organized by UNESCO in Dhaka, East Pakistan. There I became acquainted with many renowned geographers.

In 1966, at the request of the ECAFE Mekong Committee, led by Hiroshi Hori, I studied the Vientiane Plain in the middle reaches of the Mekong River. This study produced "A Geomorphological Survey Map of the Mekong River Basin Indicating Areas Subject to Flooding." In 1969, on the recommendation of Gilbert White, a professor at Chicago University, I attended the U.N. Inter-Regional Seminar on "Flood Damage Prevention Management," at Tbilisi, U.S.S.R., where I delivered a paper on "Flood Plain Adjustments, Restricted Agricultural Uses, Zoning and Building Codes as Damage Prevention Measures."

In 1972, I had a sabbatical year from Waseda University, which enabled me to attend the IGU congress in Canada. There, I was guided in the field by Archie Stalker, a research scientist with the Geological Survey of Canada. After that congress, I headed to the United States where I was able to observe the Mississippi River. I then headed to the Netherlands and taught photo-interpretation of flooding in Japan and Asia at the International Institute for Aero Survey and Earth Sciences (ITC) in Enschede.

That same year, I was appointed a member of the standing committee on Geomorphological Survey and Mapping of the IGU. From then on, I attended the annual commission meetings in Europe and was able to exchange information with many experts on geomorphological land classification, including Mieczyslaw Klimaszewski (Poland), Jaromir Demek (C.S.S.R.), Clifford Embleton (U.K.), Giovanni Castiglioni (Italy), Irene Heyse (Belgium) and Herman Verstappen (The Netherlands).

In 1976, I received a Ph.D. in science from Tokyo Metropolitan University. The title of my thesis was "Comparative Study of the Fluvial Plain Based on Geomorphologic Land Classification."

That year, at the request of the government of Bangladesh, the Japanese government organized a Jamuna River Bridge Construction Project Team. Seichi Sato, team leader for the river research, found that shifting of the river was both significant and frequent, and stabilizing its course, as is done in Japan, was impossible due to the river's large size.

As described in the present publication, my geomorphological studies led me to propose a site for the bridge at Sirajganj. My proposal was adopted, and 20 years later a bridge was constructed at that site.

In 1983, at the request of Dr. Sato, I studied the flooding and coastal erosion in the city of Padang in Sumatra by utilizing geomorphological land classification.

The Geomorphology of River and Coastal Plains Working Group of the Japan Geography Society was established that year. I chaired it, and with the support of Masatami Nakayama, Isao Takagi, Shigeko Haruyama, Sumiko Kubo and Yukihiro Hirai, we published a special issue, "River and Coastal Plains in Japan," in 1989.

I also chaired the Natural Hazards and Environmental Geomorphology Working Group of Japanese Geographers and was a member of the IGU Rapid Geomorphological Hazards Working Group.

In 1987, Dr. Haruyama and I, supported by the National Disaster prevention Center (National Research Institute for Earth Science and Disaster Prevention), began preparing a geomorphological land classification map for the Central Plain in Thailand.

In 1991, I organized the International Symposium on Flood Disaster Reduction in Southeast Asia at Waseda University. Many distinguished delegates came to it, not only from Southeast Asian countries but also from Europe.

In 1993, shortly before my retirement from Waseda University, I organized an international congress on "Geomorphological Hazards in the Asia-Pacific Region" at the university's International Conference Center. With support from Waseda University, I invited the following eminent delegates from all over the world:

Clifford EMBLETON (King's College, U.K.) and Hamann EMBLETON (Wien University)
Jan SZUPRYCSYNSZNSKI (Polish Academy of Science)
Jaromir DEMEK (Palacky University, C.S.S.R)
Jui Chin CHANG (National Taiwan Normal University)

Zhi Ming CHEN (Academy of Science of Nanjing, China)
Nutalaya PRINIYA (Asian Institute Technology, Thailand)
Charles ROSENFELD (Oregon State University, United States)
Darmowasito SUTIKNO (Gajah Mada University, Indonesia)
Irene HEYSE (State University of Ghent, Belgium)
Many Japanese experts also attended.

The results of the congress proceedings were published in Geojournal, vol. 38, 1996, as a special issue (Chief Editor Wolf Tietze). After viewing that issue, Kluwer Publishing Co. proposed that I prepare a book related to natural hazards. Hence this present publication.

There are many natural hazards such as floods, landslides, volcanoes and earthquakes in the Asia-Pacific Region. However, the study of such natural hazards in the region is poorly developed, and knowledge about their mitigation is greatly needed.

Many people used to consider that countermeasures to natural hazards were only the domain of civil engineers. However, we must concern ourselves with both structural and nonstructural methods, comprising geomorphological, geological and geographical approaches. Furthermore, there are distinct regional differences in natural hazards to which we must also pay attention.

The results of the discussions in this book apply not only to Japan and Southeast Asian countries, but also to Europe, North America and Africa.

As described above, I was supported in the making of the geomorphological maps by the Science and Technology Agency and the Ministry of Construction in Japan and by organizations such as ECAFE and JICA in other countries. I express my heartfelt thanks to all these groups.

Margaret Stalker, an editor at Asahi Evening News in Tokyo, helped with the English in this book, and her father, Dr. Stalker was of great assistance. I was supported on the sketches and preparation of the text by Yayoi Koizumi and by Hiroshi Hotta. Finally, I would like to give special thanks to my dear wife Mariko, who has so well supported my career as a geographer.

May 1st, 2000 (my 77th birthday).

OYA MASAHIKO

CHAPTER 1

GEOMORPHOLOGICAL SURVEY MAPS SHOWING CLASSIFICATION OF FLOOD-STRICKEN AREAS

1.1. History of Geomorphological Survey Mapping in Japan

After World War II, Japan faced both food shortages and flooding problems. The staple food was rice, raised mainly in alluvial plains that were frequently ravaged by flooding from typhoons and heavy rain. To increase rice production, the country needed to protect these fields.

The Japanese archipelago has experienced a great number of large-scale typhoons. Among them were the Makurazaki Typhoon in 1945, Typhoon Catherine in 1947, Typhoon Ion in 1948, Typhoon Jane in 1950, Typhoon Ruth in 1951 and Toyamaru Typhoon in 1954 (*Figure 1.1*).

Typhoon Catherine was especially devastating. The embankment of the Tone River, which has the largest catchment area in Japan, collapsed, and the main part of the Kanto Plain and the city of Tokyo (*Figure 1.2*) were inundated. The maximum flood discharge at the time was 15,000 cubic meters per second at Yattajima Research Point, near the town of Kawamata.

According to flood discharge records for the lower part of the Tone River, kept since the Meiji Restoration (1868), the levels of maximum flood discharge have gradually been increasing despite little change in the amounts of precipitation in the upper reaches of the river. Koichi Aki (1952) attributes this phenomenon in part to river conservation projects and in part to changes in land use. Civil engineers require an integrated knowledge of a river basin for flood control purposes (Oya, 1972).

During Typhoon Catherine, widespread breaches occurred on the right embankment of the upper reaches of the Tone River about four kilometers from the town of Kurihashi. The water then flowed into the Furu-Tone River (*Figure 1.3*), which was the main course of the Tone River during the Edo Period (*appended Table 2*), and stagnated in all the back-marshes (Oya and Haruyama, 1987).

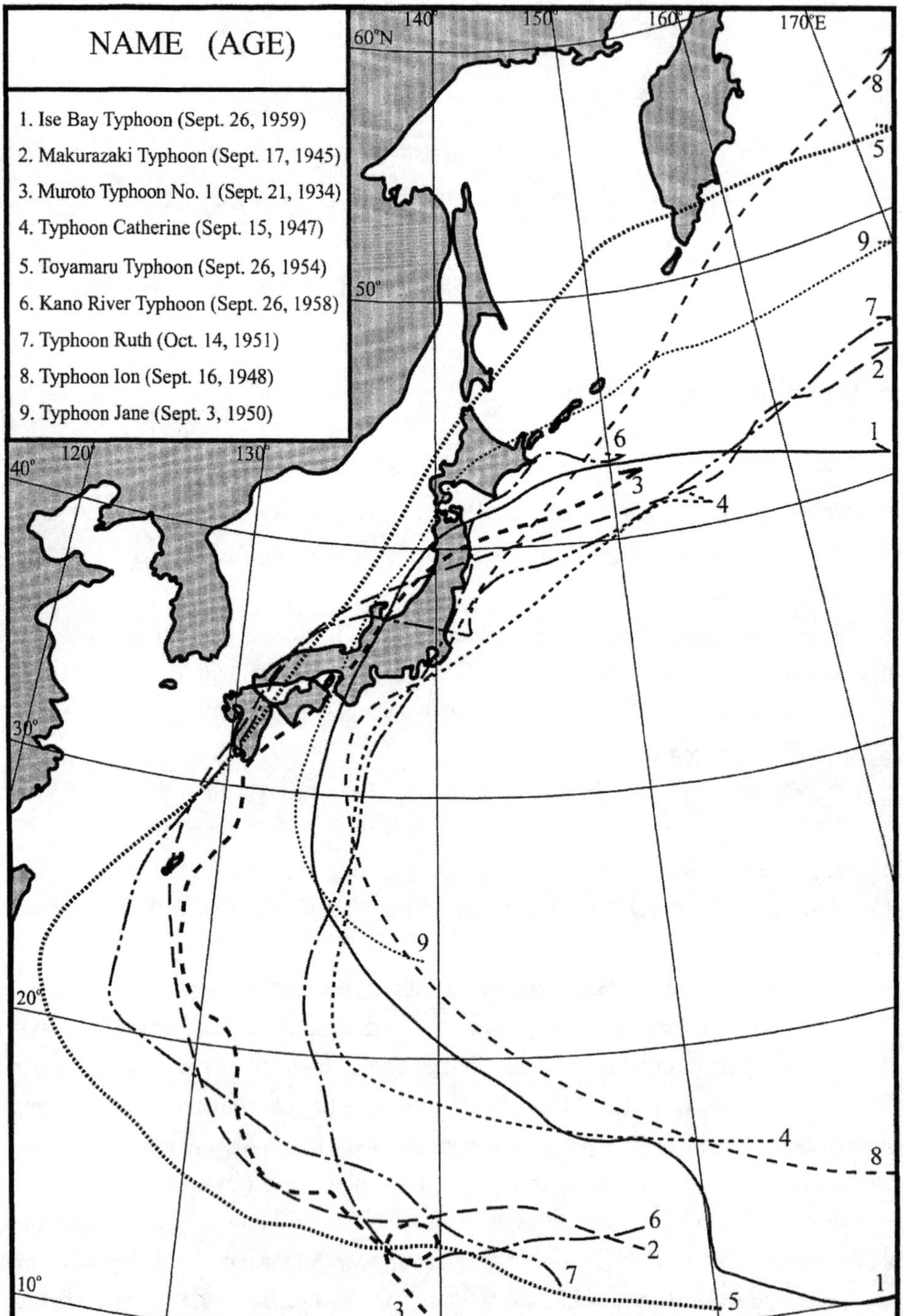

Figure 1.1 Typhoon routes and dates (Chronological Science Tables, 1989; modified)

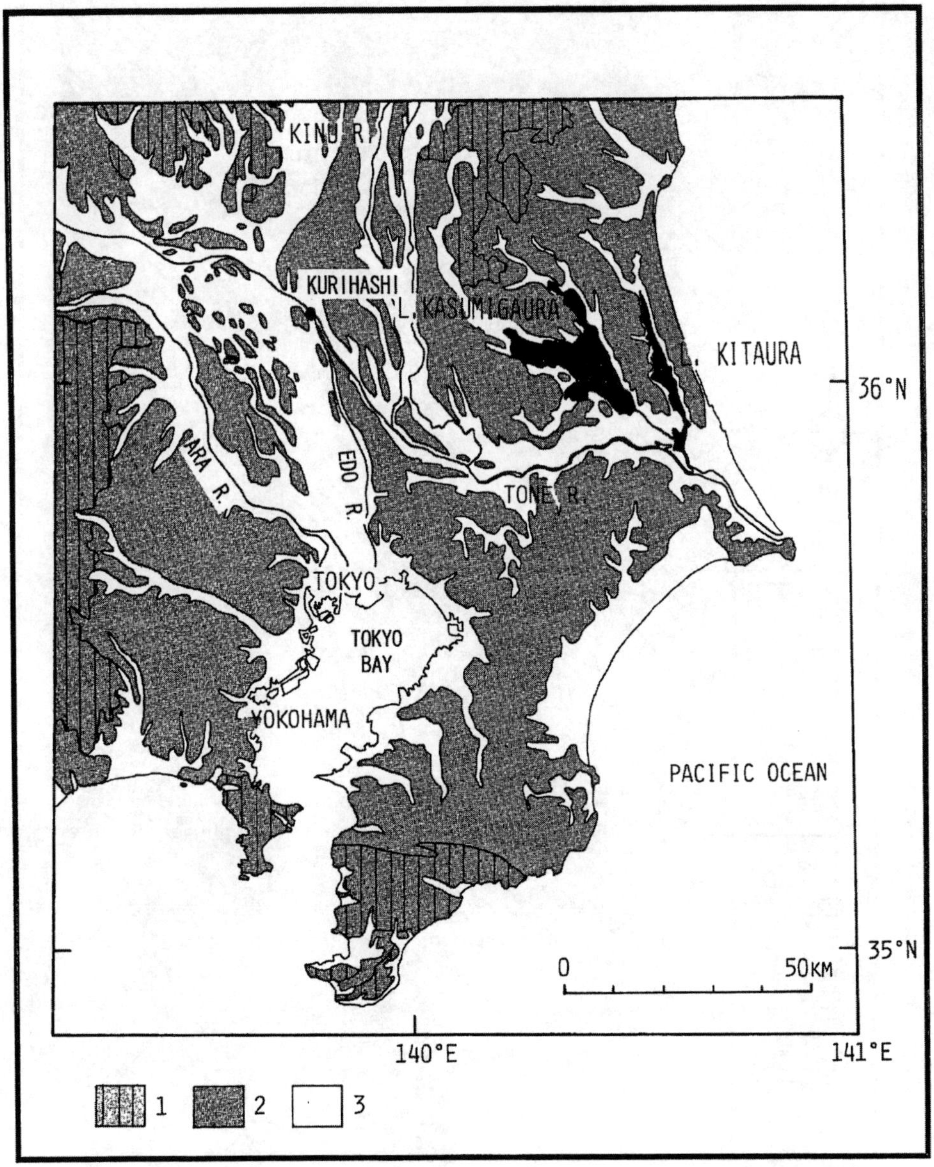

Figure 1.2 Topography of the Kanto district (Ōya, Haruyama and Hirai, 1989a; modified) (1) Mountain, (2) Hill and upland, (3) Lowland.

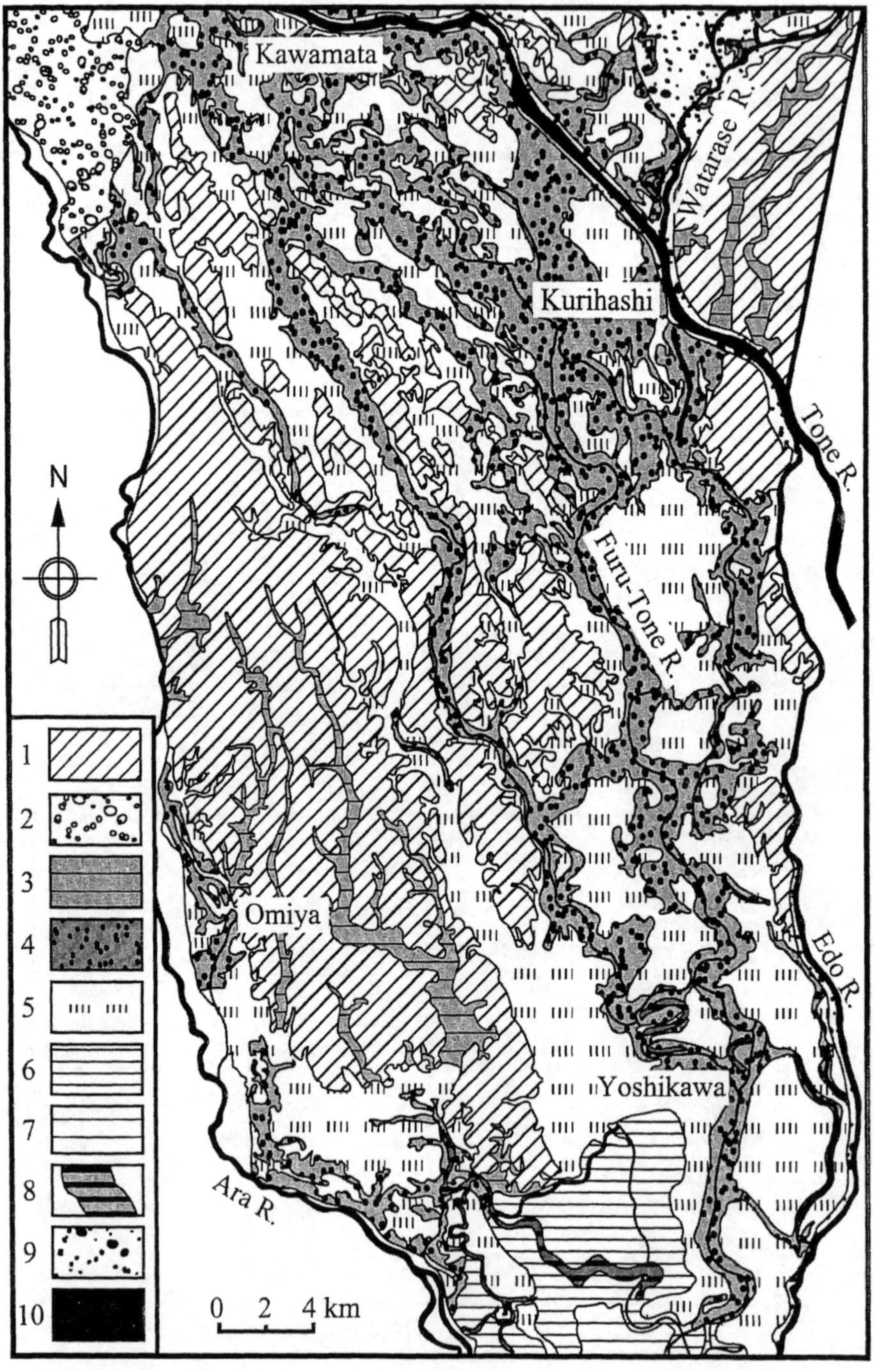

Figure 1.3 Geomorphological land classification map of a section of the Naka River basin (Oya, Ono and Kawashima, 1961; modified)
(1) Terrace, (2) Fan, (3) Valley plain, (4) Natural levee or sand dune, (5) Back-marsh and peaty area, (6) Upper delta, (7) Lower delta, (8) Former river course, (9) Dry riverbed, (10) River.

Immediately after the typhoon, the Geographical Survey Institute, Ministry of Home Affairs (1947), researched the flooding and produced several maps detailing the areas of inundation, length and depths of submergence, flood isochronal lines, and direction of current.

The research, which showed a clear relationship between the micro-topography of the plain and the features of inundation, influenced the creation of geomorphological survey maps classifying flood-stricken areas.

At the time, the Ministry of Agriculture was trying to boost grain production to combat the severe food shortages. It looked at increasing the production of chemical fertilizers but almost all such factories—having temporarily been used to manufacture gunpowder—had been bombed during the war.

There was also a proposal to improve the breeding of rice plants. However, observing the immediate need and the time it would take to put the idea into practical use, the suggestion was discarded. Only the improvement of the land itself appeared to be a feasible measure, and it required agricultural civil engineers to have a geomorphological knowledge of alluvial plains.

Fortunately, the time had already come for the study of alluvial plains among Japanese geographers. Before the war, topographic study was strongly influenced by studies in Europe and the United States. Almost all the studies had been done in mountainous regions formed by erosion, as in Europe and the United States. Furthermore, considerable parts of the plains in Europe and North America were formed by glacial processes. In contrast, the alluvial plains in Japan were formed by fluvial deposition.

Fumio Tada, who was vice president of the International Geographical Union (IGU), and his study group, including the author, sought to research the topography of depositional areas in Japan.

Immediately after World War II, the Resources Council was established by the government of Japan following the recommendation of Edward Ackerman, a geographer at General Headquarters (GHQ). The Council provided an ideal environment for discussion among civil engineers, agricultural engineers and geographers.

The Reconnaissance Topographical Survey Map of the Lower Part of the Kiso River Basin (Oya, 1956; *appended map*)—the first geomorphological survey map showing classification areas—was made under these circumstances.

1.2. Value of Geomorphological Survey Maps

A geomorphological survey map enables us to estimate the nature and extent of a submerged area, the length of time an area would be underwater, the direction of flood currents, changes in the river course, the possibility of erosion and deposition, and numerous other details.

The reason why such a survey map helps to indicate flood types is that the relief features of a plain and its sandy and gravelly deposits have been formed by repeated floods. Thus, a plain's micro-topography preserves the history of past floods.

Geomorphic features such as terraces, valley plains, fans, natural levees, back-marshes and deltas influence the extent and the nature of a flood. For example, on alluvial fans, erosion and deposition of sand and gravel are common, changes in the river channel are frequent and floodwater drains off quickly. On natural levees, the stagnant water is shallow, little sand is deposited and the water drains off well. In back-marshes, however, the water is generally deep and remains stationary for a long period, leaving a mantle of silt and clay. In a delta, the water is shallower than in a back-marsh area, but delta areas are sometimes subjected to tsunami, which are abnormal waves resulting from earthquakes, as well as tidal waves caused by typhoons (Tada and Oya, 1959).

This shows clearly that by classifying the geomorphological configuration of an area subjected to frequent flooding, one can determine the nature of previous floods as well as the probable nature of future floods. From this point of view, the author has prepared geomorphological survey maps showing classification of flood-stricken areas of the main rivers in Japan and Southeast Asia.

These maps include those of the Kiso and the Chikugo rivers in Japan, the Mekong River on the Indo-China Peninsula, the Brahmaputra-Jamuna River in Bangladesh, Padang in West Sumatra, Indonesia, and the Central Plain of Thailand. These maps were requested by the Resources Bureau of the Science Technology Agency; the Ministry of Construction; the U.N. Economic Commission of Asia and Far East (ECAFE), now called the Economic Commission of Asia and the Pacific (ESCAP); the Japan International Cooperation Agency (JICA); and the National Research Center for Disaster Prevention (*Table 1.1, appended Table 1*).

Area	Year Published
Kiso River	1956
Chikugo River	1957
Isahaya	1959
Ishikari River	1959
Naka River	1961
Sanriku Coast	1961
Yoshino River	1963
Ariake Coast	1963
Kano River	1966
Kuzuryu River	1968
Nakatsu River	1971
Neya River	1972
Tsugaru Plain	1977
Shizuoka Plain	1977
Yahagi River	1977
Shonai River	1979

Area	Year Published
Ogawara Lake	1982
Shonai Plain	1982
Abashiri Rver	1984
Agano River	1984
Kise River	1985
Tokoro River	1985
Katsushika Ward	1985
Kasumigaura, Kitaura Lakes	1986
Sagami Bay	1991
Yodo River	1993
Naruto Straits	1994
Yoshino River	1995
Ara River	1996
Naka River	1997

Area	Year Published
Mune River (Thailand)	1961
Gum River (Thailand)	1961
Mekong River and Vientiane Plain (Laos and Thailand)	1967
Brahmaputra- Jamuna River (Bangladesh)	1976
Ganges Plain (Bangladesh)	1976
Padang Plain (Sumatra, Indonesia)	1983
Nile River (Egypt)	1986
Chao Phraya River (Thailand)	1989
Kraseio River (Thailand)	1991
Atlas	1993

Table 1.1 Geomorphological Survey Maps

1.3. Method of Preparing Geomorphological Survey Maps for Alluvial Plains

To classify an area, aerial photographs must first be collected. Preferably, the photos will be of a larger scale than that of the map. For example, if the map scale is 1:50,000, the scale of the photograph should be 1:40,000.

A couple of years after World War II, all of Japan was photographed by the U.S. Army Map Service, Far East. The scale was about 1:43,000. The more important plains were later photographed by the Geographical Survey Institute of the Ministry of Construction, and mountainous regions by the Forestry Agency.

In developing countries, it is often difficult to obtain such photographs. In these cases, landsat images may be substituted. Maps should be collected chronologically.

Geomorphological land classification is easier to determine from older aerial photographs, which are less cluttered with urbanization and artificial work.

Secondly, as many topical maps as possible should be collected, including those displaying soil, geology, vegetation and land use.

In the preparation of a map, the target area is first classified into major geomorphological units such as mountains, terraces and lowlands.

Subsequently, mountains are classified by altitude, structure, form, gentle slopes on the mountain ridge, mountain flanks, piedmont gentle slopes and steep slopes. Volcanoes must also be classified by craters, mudflow, flow mound, volcanic fan, etc.

Terraces are classified into lava plateaus, pyroclastic plateaus, and upper, middle and lower terraces, based on such factors as their altitude, structural form and continuity.

Lowlands are classified into dissected fans, alluvial fans, natural levees, backmarshes and deltas. Man-made topography, such as reclaimed land and artificial fill, is also classified.

Regulations have been established for the classification of geomorphology. As well, there are several other standards of classification.

Following the creation of a geomorphological land classification map from photographs and maps, it must be confirmed by actual fieldwork.

The map, thus prepared, is then checked using information related not only to geomorphology but also flooding. The relationships between geomorphology and features of flooding are shown in *Table 1.2*.

Geomorphic Elements	State of Flooding
Terrace	Remains unsubmerged during floods.
Valley plain	When submerged, high velocity of current. Good drainage.
Fan	When submerged, erosion, deposition and change in watercourse occurs. Good drainage.
Natural levee	Submerged during extraordinary floods. Good drainage.
Back-marsh	Remains submerged for long period. Deep water.
Delta	Submerged for very long period. Experiences frequent tidal waves.
Former river course	Water flows along it during periods of extraordinary flooding.
Sand spit	Submerged during extraordinary flooding. Good drainage.
Sand dune	Remains unsubmerged.
Lowland between sand dunes	Submerged during torrential rainfall.
Reclaimed land	Submerged for very long period. Experiences frequent tidal waves.
Artificially filled field in shallow sea	Experiences tidal waves and tsunami but the water drains off well.

Table 1.2 Relationships between geomorphic elements and state of flooding (Oya, 1995; modified) (including man-made topography)

CHAPTER 2

GEOMORPHOLOGICAL STUDIES IN EUROPE

2.1. Development of Geomorphological Mapping in Europe

Geomorphological mapping is a relatively young branch of geomorphology. Whereas geological mapping began to develop in the mid-19th century, the development of geomorphological mapping proper did not start until after World War II. Its development took place mostly in Europe, Japan and Canada.

Detailed geomorphological mapping in the last four decades has become, in many countries, the key method of geomorphological survey. The progress in geomorphological mapping represents a new stage in the development of geomorphology as a science and in its practical application. However, its development has not been without obstacles.

Even in countries with a long history of geomorphological mapping, a variety of methodological approaches can still be found. Maps compiled in individual countries often differ considerably. The heterogeneity of methodological approaches and means of representation have seriously impeded the development of geomorphological mapping.

At the 19th IGU Congress in Stockholm, 1960, it was decided on the initiative of Mieczyslaw Klimaszewski of Poland to establish a subcommission on geomorphological mapping within the IGU Commission on Applied Geomorphology. At the IGU Congress in New Delhi, 1968, the subcommission was dissolved into a new commission, Geomorphological Survey and Mapping. After that, numerous meetings were held in European countries for the preparation of the International Geomorphologic Map of Europe.

The commission was guided by the following European members: Jaromir Demek (C.S.S.R.), Mieczyslaw Klimaszewski (Poland), Leszek Starkel (Poland), Clifford Embleton (U.K.), F. Joly (France), Jean Tricart (France), Johannes Gellert (Germany), H. Kugler (Germany), E. Scholz (Germany), Giovanni Castiglioni (Italy), N. Bauer (Austria), Marton Pecsi (Hungary), F. Gullentops (Belgium), Irene Heyse (Belgium),

Maria Sala (Spain), N. V. Bashenina (U.S.S.R.) and N. V. Demitrashko (U.S.S.R.). Furthermore, Denis St. Onge (Canada), Gordon Wolmann (U.S.), Mabutt (Australia) and Masahiko Oya (Japan) participated in the commission.

The legends for the map were compiled by Demek into the "Manual of Detailed Geomorphological Mapping" (1972). With the assistance of UNESCO and C.S.S.R. Academy of Sciences, the commission completed in 1976 the International Geomorphological Map of Europe, which was displayed on 16 sheets with a scale of 1:250,000. A book titled "Geomorphology of Europe" (1984), compiled by Embleton, was based on this map. The book not only discussed the international geomorphological map but also included the most recent knowledge on the geomorphology of Europe. The book was translated into Japanese, "Yoroppa no Chikei" (1997), by Oya, Yukiyasu Saka and members of the Geomorphology of River and Coastal Plains Working Group of the Association of Japanese Geographers.

2.2. Geomorphological Mapping Working Groups in Europe and Japan

The Geomorphology of River and Coastal Plains Working Group was set up at the 1980 IGU Congress in Tokyo. The group was guided by Joop ten Cate (The Netherlands), Oya (Japan), Castiglioni (Italy) and Jan Szuprycznski (Poland).

The purposes of the working group were: a) to compile geomorphological maps of flood areas in some of the world's most important plains and b) to make a report on the geomorphology of plains related to natural and man-made hazards (Oya, 1990).

Meanwhile in Japan, the Geomorphology of River and Coastal Plains Working Group of the Association of Japanese Geographers was founded in 1979 and presented 85 studies from 1979 to 1990. The group was guided by Masatami Nakayama, Isao Takagi, Shigeko Haruyama, Sumiko Kubo and the author. The main features of these studies can be summarized as follows:

a) Almost all the studies were based on geomorphological land classification.

b) Members focused not only on the geomorphology of alluvial plains but also on the overall condition of drainage basins. Most notable was the strong influence of an intermontane depression in the upper reaches on an alluvial plain in the lower reaches.

c) Members examined the relationship between the micro-topography of the plains and archaeological features such as shell mounds, which indicate the location of former coastlines.

d) Applied geographical studies were carried out using geomorphological land classification maps. The maps were used in the identification of areas subject to inundation and soil liquefaction due to earthquakes.
e) Integrated regional studies were carried out in contrasting climatic zones such as the Okhotsk Sea coastal plain in the semi-frigid zone and the outer zone of southwest Japan, which is in the humid temperate zone.
f) Members compared alluvial plain research carried out in the stable continental zones of other Asian regions such the Korean Peninsula and the Central Plain of Thailand with that of Japan (Oya, 1989b).

2.3. The International Geomorphological Map of Europe

The International Geomorphological Map of Europe incorporates four major elements in its conceptional framework: a) the differentiation of relief classes according to relief amplitude, b) the differentiation of relief according to morphostructural type, c) the differentiation of "special relief forms" of mainly exogenous origin and d) chronological land classification (Embleton, 1984).

For relief types, five classes of relief amplitude are used: faint, low, moderate, mountainous and high mountain.

Relief types are schematically demonstrated in *Figure 2.1*, including predominantly erosional relief; depositional relief developed on thick sediments overlying a deep-seated base; depositional relief developed on thinner sediments overlying a shallow-seated base, involving uplift of parts of platforms; marginal or intermontane depressions; and erosional-depositional relief on a shallow-seated base.

There are altogether only 13 megamorphostructures. Erosional relief types, for example, include six megamorphostructures: relatively stable shields, relatively stable platforms, epiplatform zones slightly affected by repeated tectonics, shield zones affected by neotectonics, epiplatform zones strongly affected by neotectonics and young epigeosynclinal orogenic zones.

Volcanic relief includes only a single morphostructure: Cenozoic volcanism, which gives rise to a variety of volcanic landforms.

Depositional relief on a deep-seated base is confined to marginal and intermontane depressions and platform depression. Depositional relief on a shallow-seated base is also restricted to similar depressions and basins.

Erosional-depositional relief on a shallow-seated base is formed not only in marginal and intermontane depressions under conditions of still more active neotectonic uplift, but also on uplifted parts of the crystalline basement beneath plains, and beneath platforms and shields.

Each megamorphostructure is classified on the basis of its formation process. Such an approach reflects the principles on which the legend is based.

The topography was formed by crustal movement and by exogenous factors such as aeolian, glacial, fluvial and marine. For example, sand dunes were formed by aeolian deposition; eskers, drumlins, cirques and fiords were formed by glacial erosion and deposition; fans and natural levees were formed by fluvial deposition; and deltas were formed by marine-fluvial deposition.

While the North European Plain, stretching from the Low Countries through Germany into Poland, is associated with a single large platform depression, the depositional relief is due to deposits of very different origins: marine, glacial, glaciofluvial and fluvial.

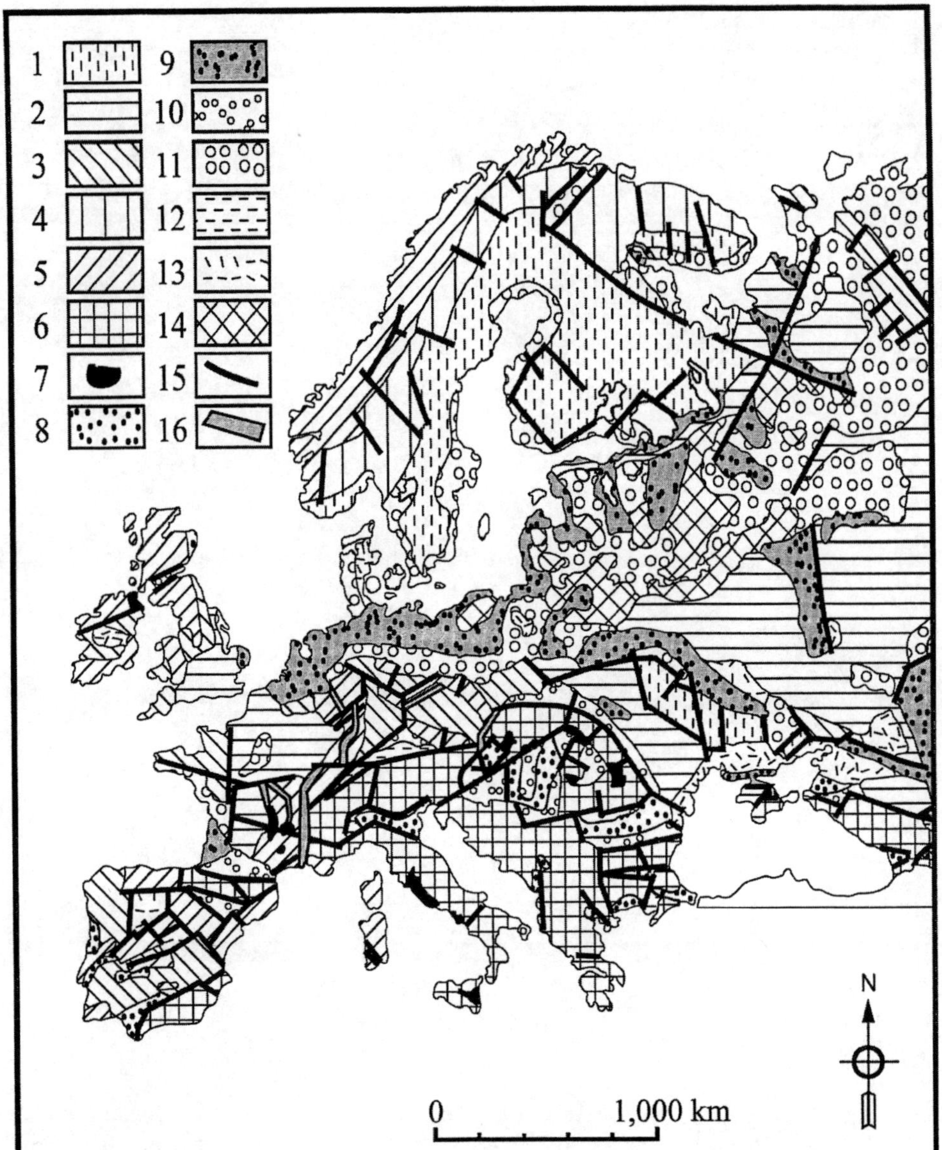

Figure 2.1 Megamorphostructures of Europe (after Bashenina et al., 1979; modified)
(A) Erosional relief: (1) Relatively stable shields, (2) Relatively stable platforms, (3) Epiplatform areas slightly affected by later tectonics, (4) Shield areas affected by neotectonics, (5) Epiplatform areas strongly affected by neotectonics, (6) Young epigeosynclinal orogenic zones.
(B) Volcanic relief: (7) Cenozoic volcanics.
(C) Depositional relief on a deep-seated basement: (8) Marginal and Intermontane depressions, (9) Platform and shield depressions.
(D) Depositional relief on a shallow basement: (10) Marginal and intermontane depressions, (11) Platform depressions.
(E) Erosional-depositional relief on a shallow basement: (12) Marginal and intermontane depressions, (13) Platform depressions, (14) Platform and shield rises, (15) Faults, (16) Rift valleys.

CHAPTER 3

VALIDATION OF A GEOMORPHOLOGICAL SURVEY MAP

3.1. Geomorphology of the Nobi Plain

The Reconnaissance Topographical Survey Map of the Lower Part of the Kiso River Basin, produced by the author in 1956, shows the Kiso, Nagara and Ibi rivers flowing in the western part of the Nobi Plain, facing Ise Bay (*appended map*). The altitude of the plain's western part is lower than that of the eastern part as a result of crustal movement. Pleistocene terraces are developed only in the eastern part.

The centers of Japan's three largest cities—Tokyo, Osaka and Nagoya—are located on terraces about 10 to 30 meters above sea level, which remain unsubmerged during floods. There are natural levees with back-marshes between the lower edge of the fan and the upper edge of the delta. These natural levees reveal the main course and branch courses of the Kiso River before the 15th century. Back-marshes occupy the spaces between the natural levees, as well as between the natural levees and fans or terraces.

If we were to draw a line between the cities of Nagoya and Tsushima, we would find it to be the border of the delta, the southernmost part of which was land reclaimed after the 17th century.

Around Nagoya Port is an artificially filled field, which was constructed after the Meiji Restoration. Other than this field, which is 1.5 to 3.5 meters above sea level, the delta is almost all at sea level. About 185.4 square kilometers of the Nobi Plain is below sea level, and of this, about 40 percent is more than 1 meter below mean sea level—the largest area to be below sea level of any plain in Japan (*Figures 3.1*, *3.2*).

3.1.1. THE ISE BAY TYPHOON

On Sept. 22, 1959, low atmospheric pressure near Saipan Island, in the Caroline Archipelago, developed rapidly into the super-scale Ise Bay Typhoon.

On Sept. 26, the typhoon hit the Kii Peninsula in central Japan, passing north of Nagoya before proceeding to the east along the coast of the Japan Sea.

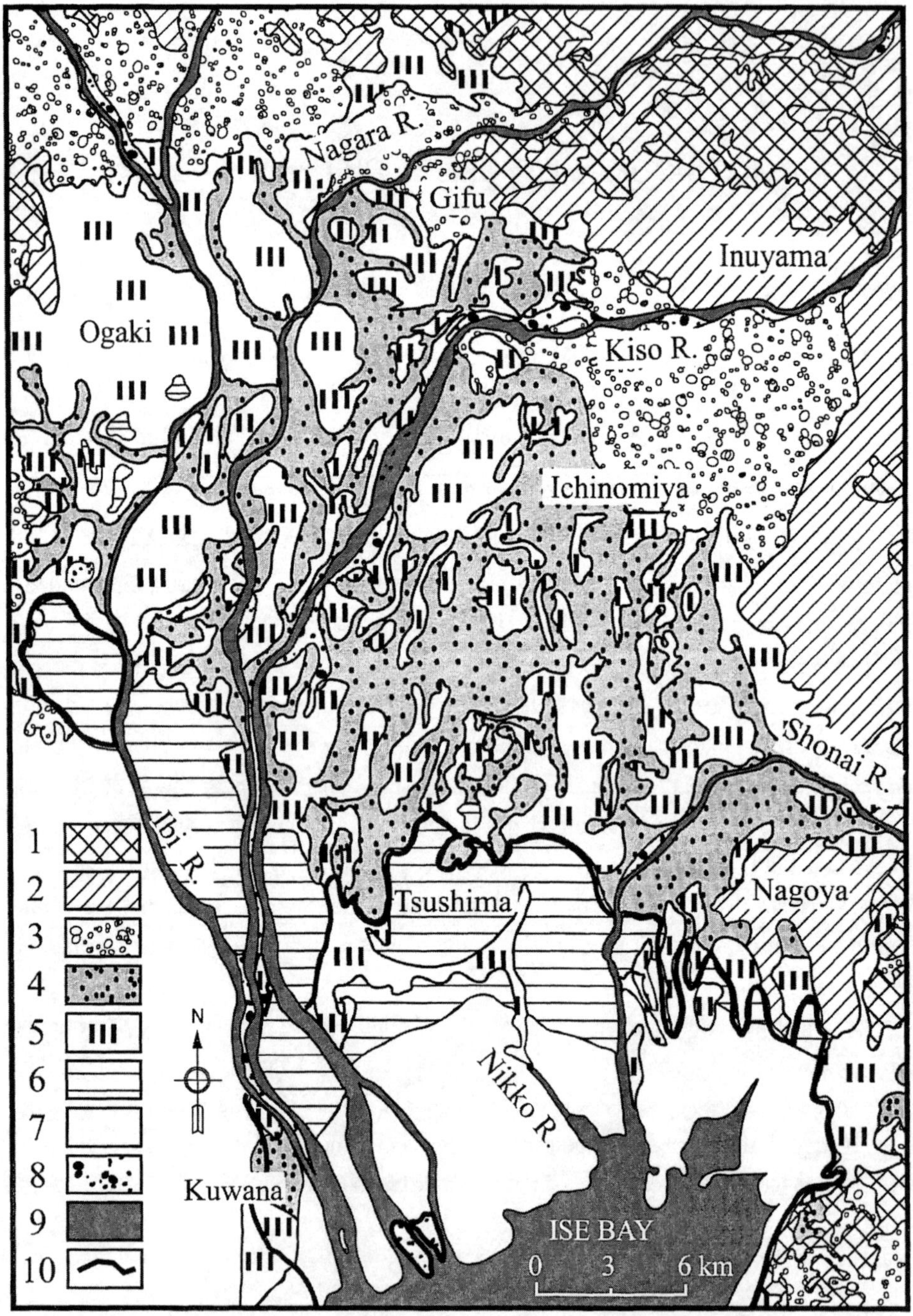

Figure 3.1 Geomorphological land classification map of the Nobi Plain (Oya, 1956; modified)
(1) Mountain, (2) Terrace, (3) Fan, (4) Natural levee, (5) Back-marsh, (6) Delta, (7) Reclaimed land, (8) Dry riverbed, (9) River or sea, (10) Limit of inundated area following the Ise Bay Typhoon.

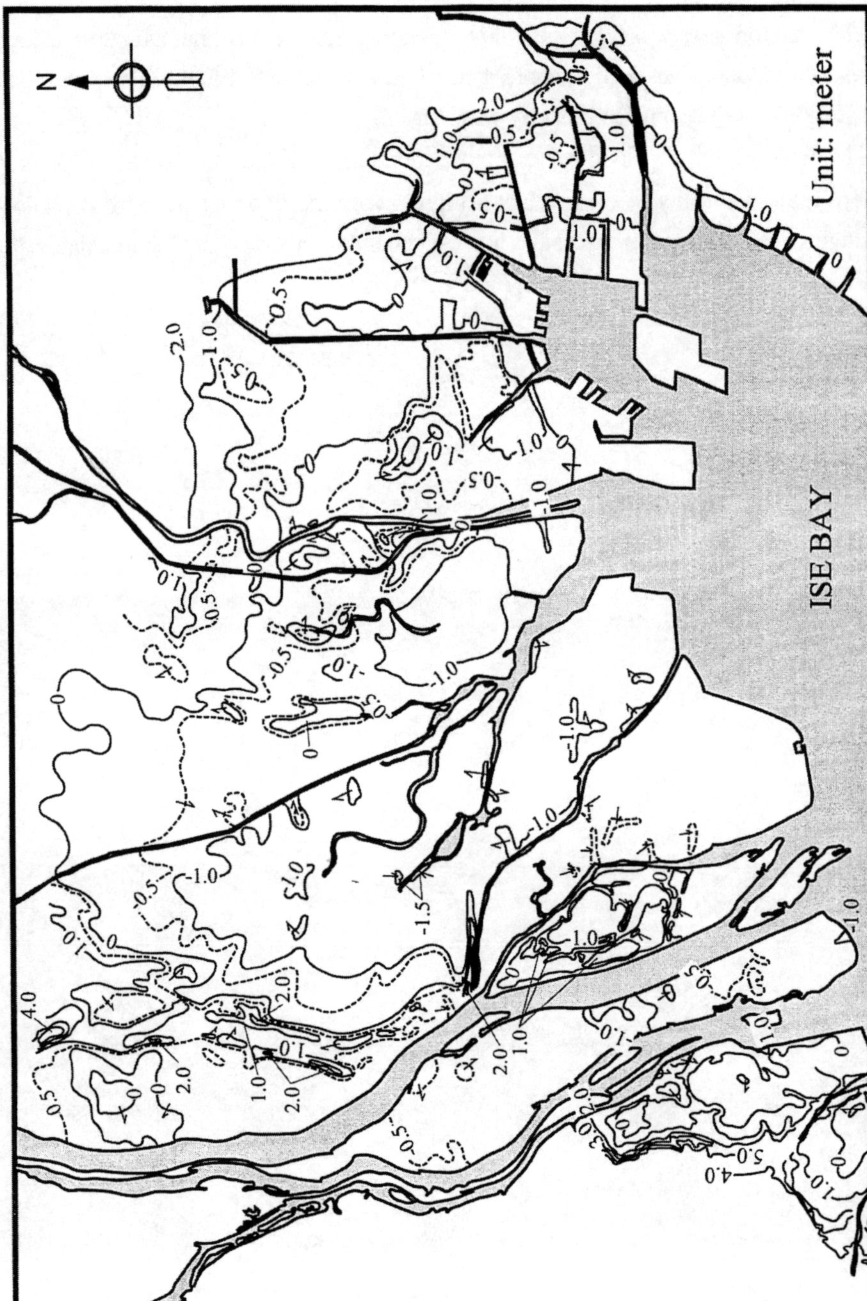

Figure 3.2 Altitudes in the Ise Bay area (Geographical Survey Institute, 1960; modified)

In Nagoya, the mean velocity of the wind was 37 meters per second and the instantaneous velocity of the wind was 46 meters per second. The atmospheric pressure was 929.5 hectopascal.

The typhoon caused a tidal wave that was reported to be as high as 3.89 meters at Nagoya Port. More than 5,200 people died tragically and about 530 billion yen (about $1.472 billion at that time) in property was lost.

The features of the flood were almost identical to those predicted by the map, demonstrating its utility. A close relationship between the flooding and geomorphology was highly visible in many cases, and it is especially noteworthy that the inundated area coincided exactly with the delta area (*Figure 3.1*).

Photo 3.1 The city of Gifu is situated on the fan of the Nagara River. (1994)

Photo 3.2 A persimmon orchard covers a natural levee in the Nobi Plain
The back-marsh is used as a paddy field. (1978)

Photo 3.3 "Waju" dike (circular levee) in the western part of the Nobi Plain (1994)

Photo 3.4 The lower reaches of the Kiso River, left, and Nagara River, right, in the Nobi Plain (1994)

Photo 3.5 Delta and a forested natural levee at the city of Tsushima in the Nobi Plain (1994)

3.1.2. FEATURES OF FLOODING

Flood features, such as the extremities of inundation, the depth of the stagnant water, its period of stagnation, the velocity of the flood current and the levels of erosion and deposition, vary remarkably according to a region's landforms. Fans, natural levees, back-marshes and deltas have a decisive influence upon determining the types of floods (Nakano et al., 1960).

Invasion Routes of Tidal Waves

The direction a giant wave will take is determined by landforms such as rivers, channels and ancient waterways.

a) Rivers: The Kiso, Nagara and Ibi rivers, along with some smaller waterways, made accessible routes for the tidal wave during the Ise Bay Typhoon. The wave, with a velocity ranging from 5 to 8 meters per second, proceeded about 24 kilometers from the mouth of the Kiso River.

b) Channels: Nagoya has many channels, including Nakagawa and Horikawa. However, the gates at the mouth of the Nakagawa Channel prevented the wave's intrusion. After the typhoon, gates were also installed at the mouth of the Horikawa Channel.

c) Former river courses: Converted into paddy fields several hundred years before the Ise Bay Typhoon, these old routes were still very accessible to the high wave's intrusion.

Limits of Intrusion

After the 3.89-meter wave hit the coast of Nagoya and vicinity, the entire delta area was covered by seawater, but the water did not flow horizontally inland. To ascertain why this did not happen, the author has profiled the wave's flow.

Figure 3.3 shows the location. The author and colleagues determined the flow's course from watermarks left on the walls of buildings and debris deposited on banks.

Figure 3.4 is a longitudinal cross section of the flooded area. The darker shaded area depicts ground levels. Reclaimed land lies next to the sea and the inland areas are delta. Almost all of this region is below sea level.

The profile shows how the seawater lowered rapidly as it flowed inland. Its depth at the seashore was about 5 to 6 meters but only about a meter at a point about five to six kilometers from the coast. But beyond this point, the water became deeper once more as the regular tidal flow swept through the eroded embankments.

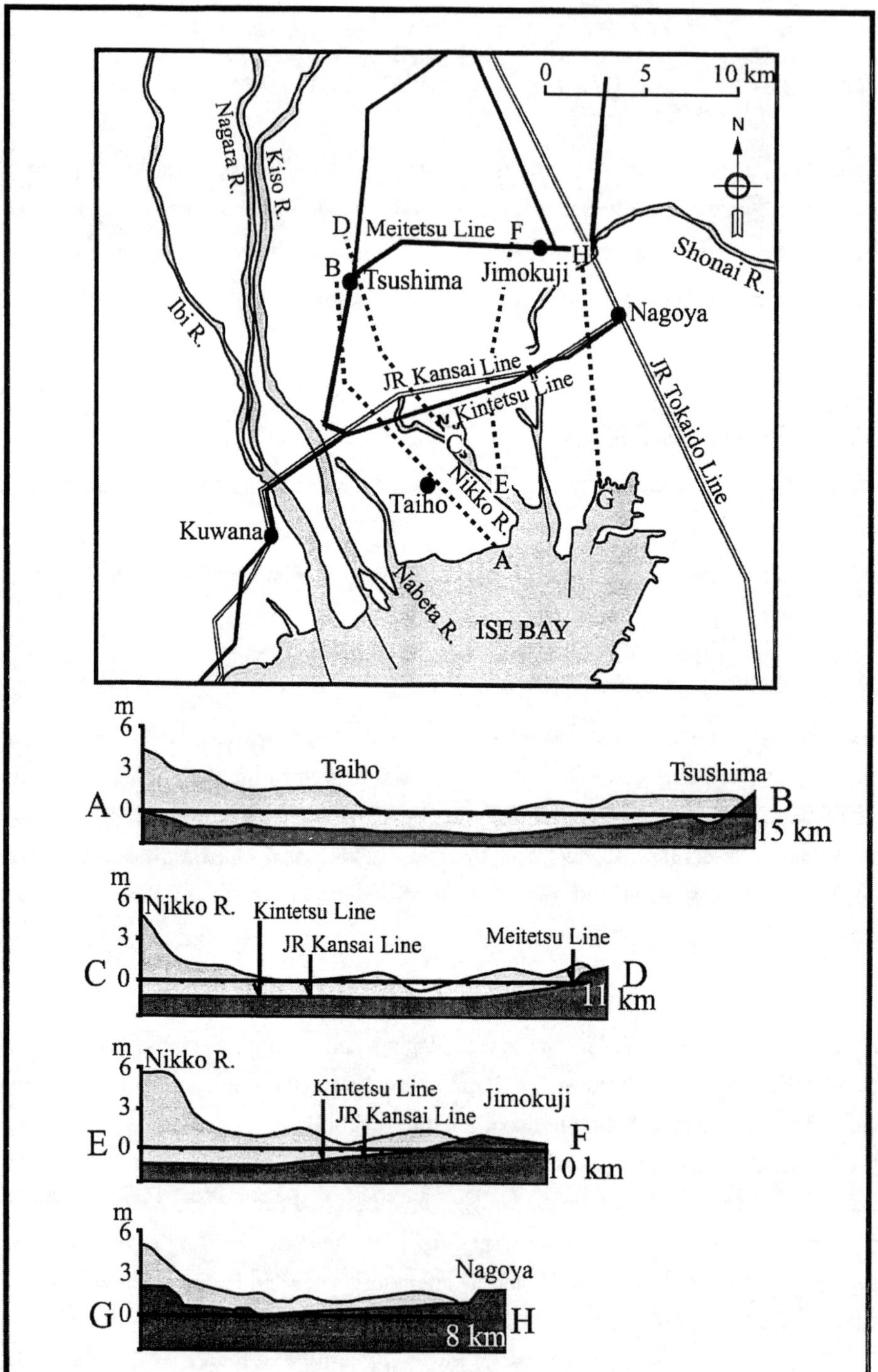

Figure 3.3, 3.4 Location and profiles of tidal wave inundation (Nakano and Oya, 1960; modified)

The wave's velocity was about 15 to 20 kilometers per hour from the coast to about this point. Beyond this, velocity decreased rapidly. The rapid lowering of the seawater was due to friction between the water and the ground surface, embankments and buildings.

The regular high tide of the following day pushed through the banks that had been destroyed. Seawater reached Tsushima, in the uppermost part of the delta, on Sept. 28, 1959.

The profile shows how the land rises at that point. This was the coastline during the 14th to 16th centuries, and became the limits of the spring tide following the Ise Bay Typhoon.

The Nobi Plain has been devastated by tidal waves 10 times since 1868. Waves on Sept. 11, 1889, and Sept. 11, 1896, were particularly high. While the flood levels resulting from the two high waves were different, their limits were the same—namely, Tsushima.

The recognition that Tsushima has been the extreme limit in a number of floods caused by typhoons is important, as it allows us to forecast tidal limits by geomorphological survey. From this perspective, the author studied the plain along the Ariake Sea in Kyushu.

3.2. Application to the Ariake Sea Lowland

At the request of the Coast and River Division of the Ministry of Construction, the author has estimated the potential extent of flooding from high waves in the Ariake Sea Lowland.

This lowland consists of fans, lower fans, upper deltas, deltas and reclaimed land. Whereas most rivers in the area flow from north to south, the Sagaego River, situated east of the city of Saga, runs from west to east, i.e., parallel with the coastline. Many shell mounds have been discovered in the northern part of the lowland but none in the southern part. The ground level of the northern part is more than 3 meters above sea level but is less than that it the southern part (*Figure 3.5*).

The author considers that an imaginary line connecting Saga with Sagaego River shows a former coastline, correlating with the line connecting Nagoya with Tsushima in the Nobi Plain.

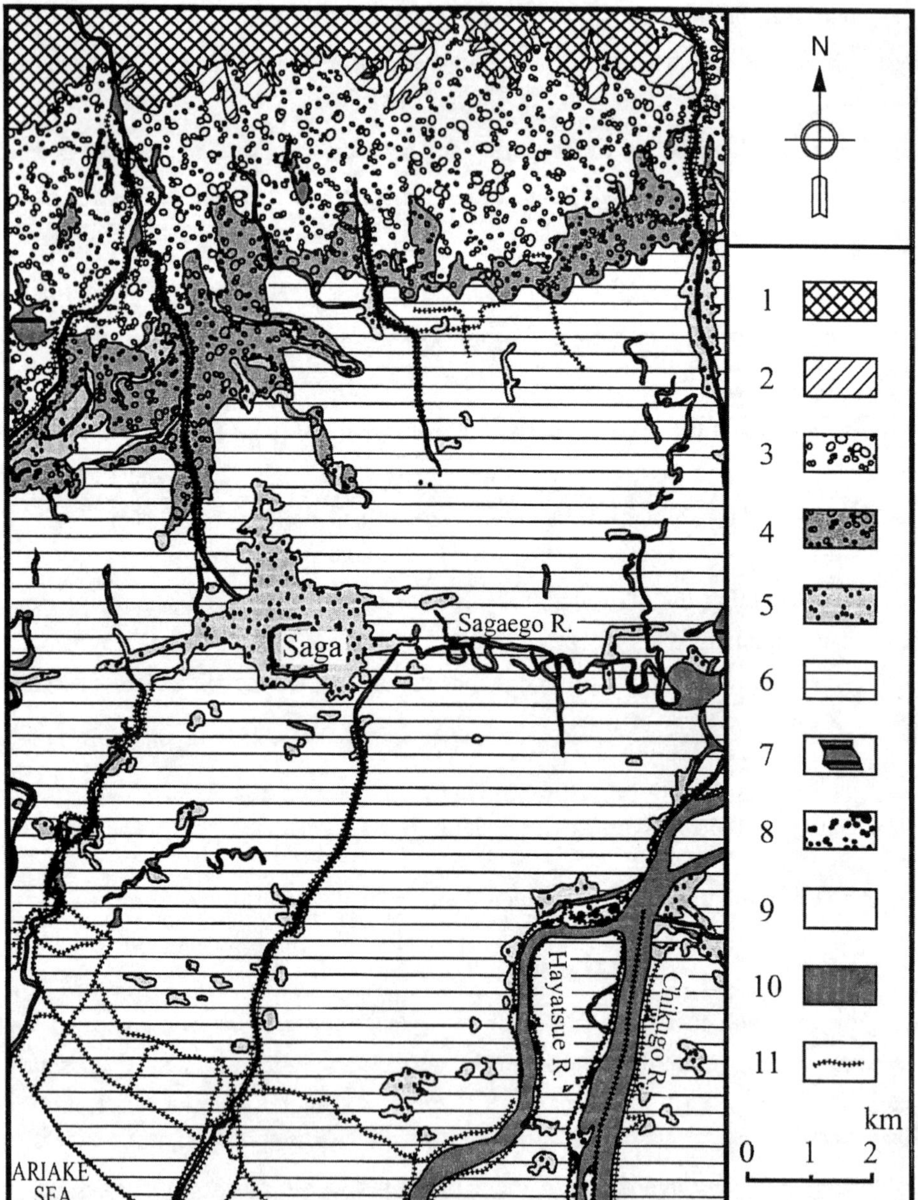

Figure 3.5 Topographical survey map of the lower reaches of the Chikugo River (Oya, 1957; modified) (1) Mountain, (2) Terrace, (3) Fan, (4) Lower fan, (5) Natural levee, sandbank or artificially filled field, (6) Delta, (7) Former river course, (8) Dry riverbed, (9) Tidal land, (10) River, (11) Embankment.

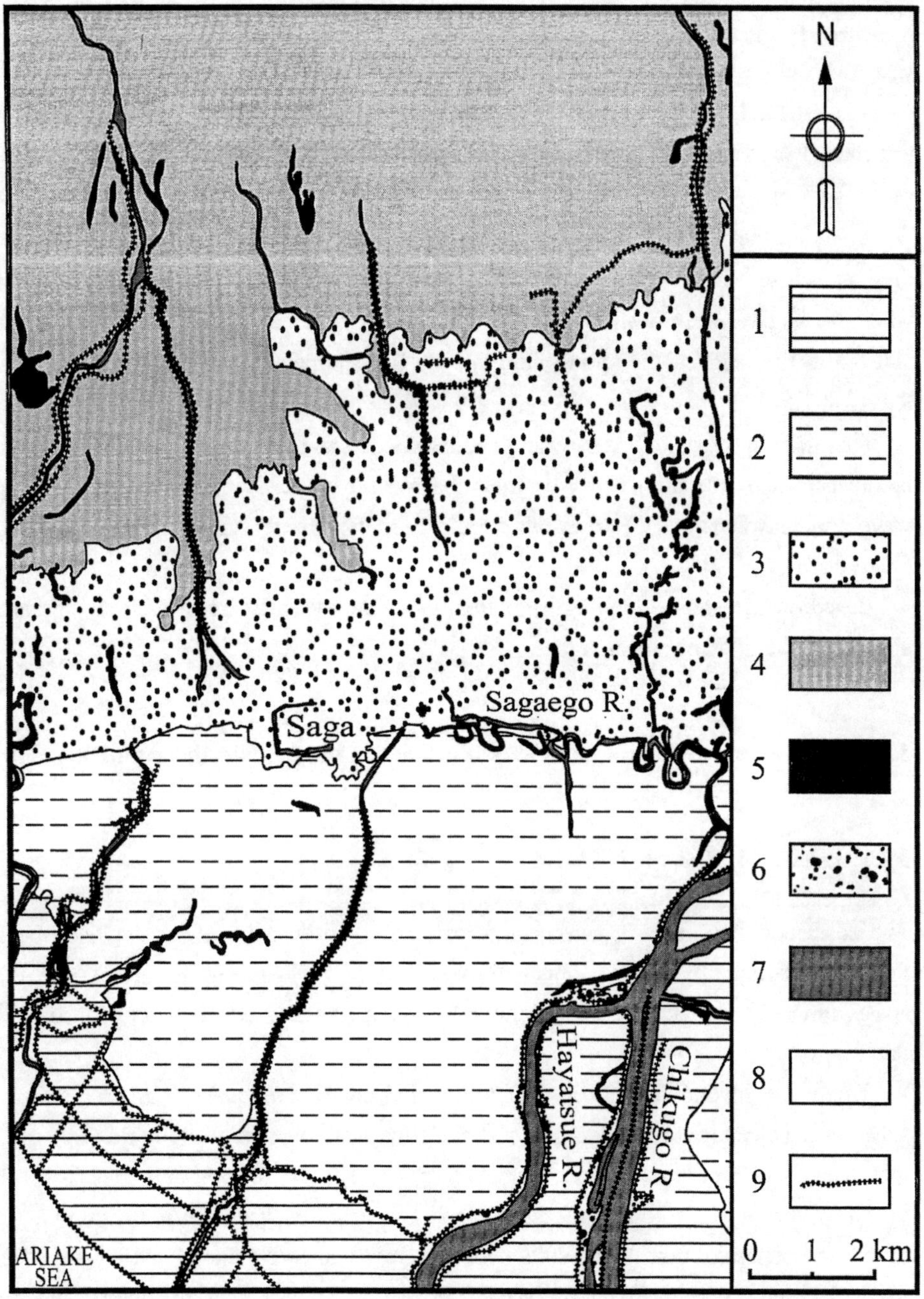

Figure 3.6 Estimation of inundated areas following a tidal wave (Oya, 1957; modified)
(1) Area that would be devastated by a tidal wave, (2) Area that would be partly devastated by a tidal wave and seawater, (3) Area that would be only slightly submerged, (4) Area that would remain unsubmerged, (5) Former river course, (6) Dry riverbed, (7) River, (8) Tidal land, (9) Embankment.

There have been several tidal waves in the area, including one resulting from a typhoon that occurred on Sept. 14, 1959. Using records from previous typhoons, the Ministry of Construction calculated the wave to be 7.5 meters high when the typhoon was at its severest stage.

It is believed that future typhoons could result in similarly high floodwater levels if the coastal embankments are destroyed by the waves. If so, almost all of the plains, including Saga, would be flooded, with great damage resulting.

Applying the phenomenon seen during the Ise Bay Typhoon, the tidal wave would dissipate close to the seacoast. However, seawater would come through the places where banks have been destroyed by the daily high tides, reaching the former coastline situated at the level of the spring tide, i.e. the line connecting Saga and the Sagaego River.

Therefore, the Ariake Sea Lowland can be divided as following: a) areas that would be devastated by the high wave, b) areas that would be partly devastated by the high wave and seawater, and c) areas that would be only slightly submerged by seawater (*Figure 3.6*).

Based on the above flooding estimates, the ministry calculated flood damage and appropriate heights for embankments.

3.3. Using Geomorphological Land Classification to Estimate Flooding from Tsunami

3.3.1. TSUNAMI IN JAPAN

The coasts of the Japanese archipelago, especially bordering the Pacific, have been frequently battered by tsunami. Since the year 684, large-scale tsunami have struck the coasts more than 50 times, with about 80 percent of these attacks occurring on the Pacific Coast.

There are several causes of tsunami, including a) earthquakes, b) eruptions of subaquatic volcanoes and c) landslides in the subaerial and subaquatic regions. Among these causes, a) is predominant in Japan, whereas b) is more frequent in Indonesia.

a) The Sanriku Coast of northeastern Japan, on the Pacific Ocean, is known for disastrous tsunami. More than 23,000 people were killed by a tsunami in 1896. Another in 1933 killed 3,000. On Ishigaki Island, between the East China Sea and the Pacific Ocean, 8,400 people drowned in a 1771 tsunami that reached heights of 30 meters.

b) About 36,000 people were killed along the Sunda Strait in 1883 by the eruption of Krakatou Volcano in Indonesia.

c) In 1792, Mount Mayuyama, located near the town of Shimabara in Kyushu, collapsed and a lot of sand and gravel fell into the Ariake Sea. This created a tsunami of more than 10 meters in height, which hit the coasts of the Shimabara Peninsula and Kumamoto Prefecture, killing 10,530 people.

The size and characteristics of tsunami formed by earthquakes near the Japanese archipelago are quite different from those originating from earthquakes far away, for example, in places such as Chile and Peru.

The scale of a tsunami is determined not only by the size of the earthquake or submarine eruption that caused it and how far its epicenter is from the coast, but also by its direction and the shape and size of the embayment.

On May 24, 1960, the tsunami from a Chilean earthquake hit the Pacific Coast of Japan. The author researched the effects of this tsunami on the Sanriku Coast and Kii Peninsula, focusing on the relationship between coastal landforms and tsunami disasters (Nakano et al., 1962).

3.3.2. COMPARING TSUNAMI FROM NEAR AND AFAR

There are distinct differences between tsunami originating locally and from far away, with the latter affecting a greater area but with a relatively calmer water surface.

When a tsunami originates nearby, high, upright waves hit the coast but the wave height decreases rapidly as it travels inward. When a tsunami originates from afar, the wave height at the entrance of a bay is lower but becomes increasingly higher.

Generally speaking, tsunami from afar are less aggressive along the coastline than tsunami originating nearby.

3.3.3. COASTAL LANDFORMS OF AFFECTED REGIONS

It is well known that the rias type of coasts along the Pacific Ocean of Japan have frequently been hit by tsunami. Chilean earthquakes have produced tsunami along the rias coastlines of the Sanriku Coast, the Kii Peninsula, southern Shikoku and southeastern Kyushu.

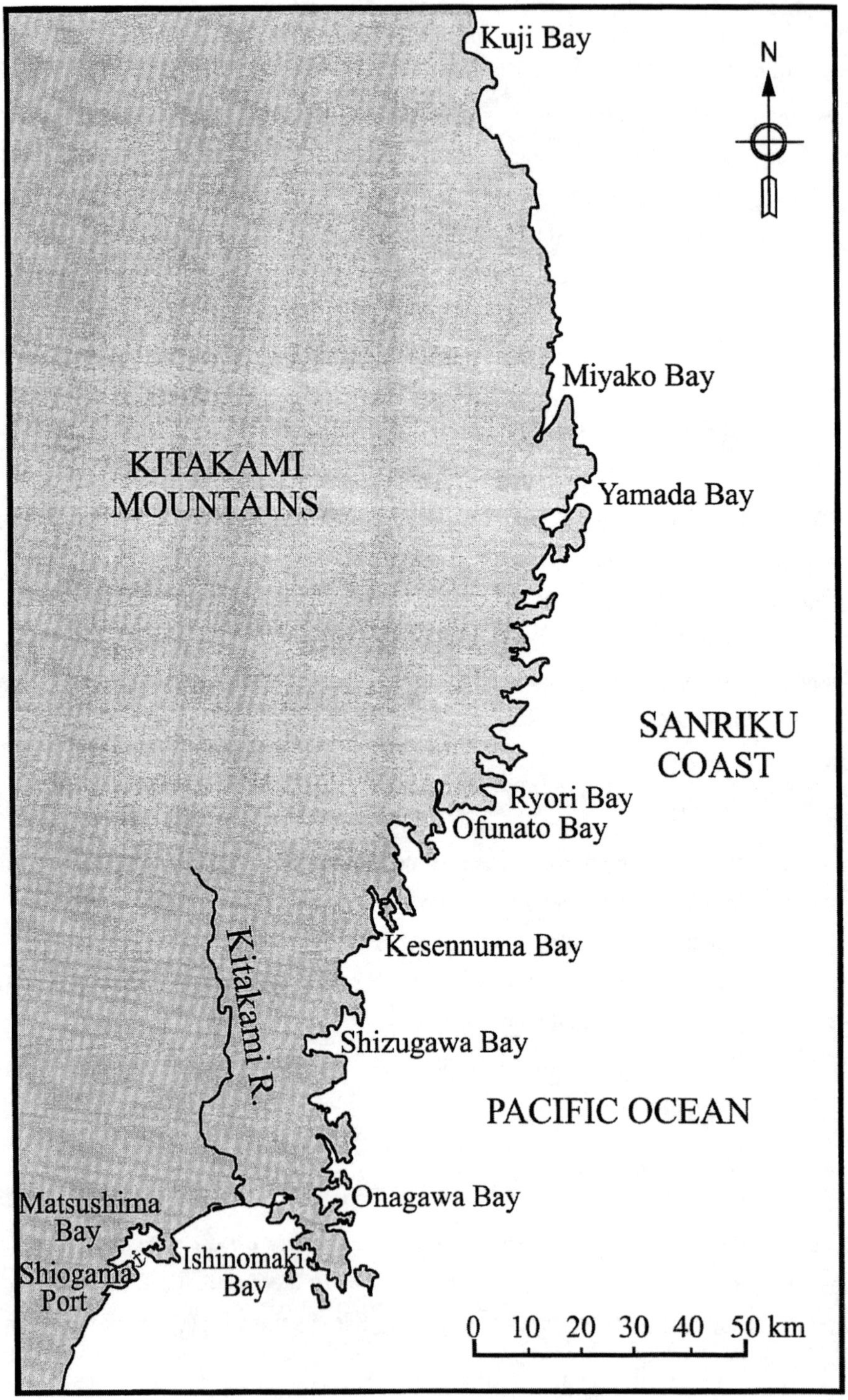

Figure 3.7 Sanriku Coast (Oya, 1962; modified)

Besides the rias-type coasts and embayments, sand bars connecting an island with a main island, such as Kiritappu in east Hokkaido, are also affected by tsunami.

Rias-type coasts or embayments are classified into several types, including U-shaped, V-shaped, W-shaped and half-moon shaped. A U-shaped embayment often has an irregular pattern, and some of the embayments are separated by barriers into several parts (*Figure 3.7*).

Typically, a U-shaped embayment was once a large valley. The submarine landform reflects the form of the former valley and tends to be rather deep. Such properties are conducive to the invasion of tsunami. Sanriku Coast and the bays of Miyako, Yamada, Ofunato and Kesennuma are typical U-shaped embayments that have all suffered tsunami disasters. In the Kii Peninsula, the U-shaped bays of Owase, Yuasa and Kuroe, which contains Kainan Harbor, have all experienced severe tsunami.

V-shaped embayments originated from rather small valleys. Sometimes two V-shaped embayments are found together, creating a W-shaped bay. Along the Sanriku Coast, Shizugawa, Ryori and Onagawa bays are typical V-shaped embayments that have experienced severe tsunami. In the Kii Peninsula, there are V-shaped bays from Kinomoto to the north, some of which are of the dendritic type.

Sand bars are easily overwhelmed by tsunami, with sometimes disastrous results. Such a disaster occurred in Kiritappu, east Hokkaido, and to a lesser extent, in Kushimoto in the Kii Peninsula. As well, sand spits and delta regions are easily inundated (*Figure 3.8*).

3.3.4. TSUNAMI ROUTES

The heights of tsunami originating from far away are diminished by the properties of embayments and submarine landforms. In an embayment, submerged valleys become the invasion routes of a tsunami unless its velocity is very great. However, the tsunami resulting from the 1960 Chilean earthquake was not fast. Its main flow entered the Sanriku Coast and Kii Peninsula embayments through recesses in the shallow water, known as *mio*. High waves from typhoons have been observed using similar paths.

Submerged valleys do not always have smooth surfaces and sometimes have irregular floors, factors that slow tsunami movement. A typical example is the tsunami that occurred in Urado Bay, Shikoku, following the Chilean earthquake. Tidal records indicate its height was probably diminished by the irregular shape of the bay's floor.

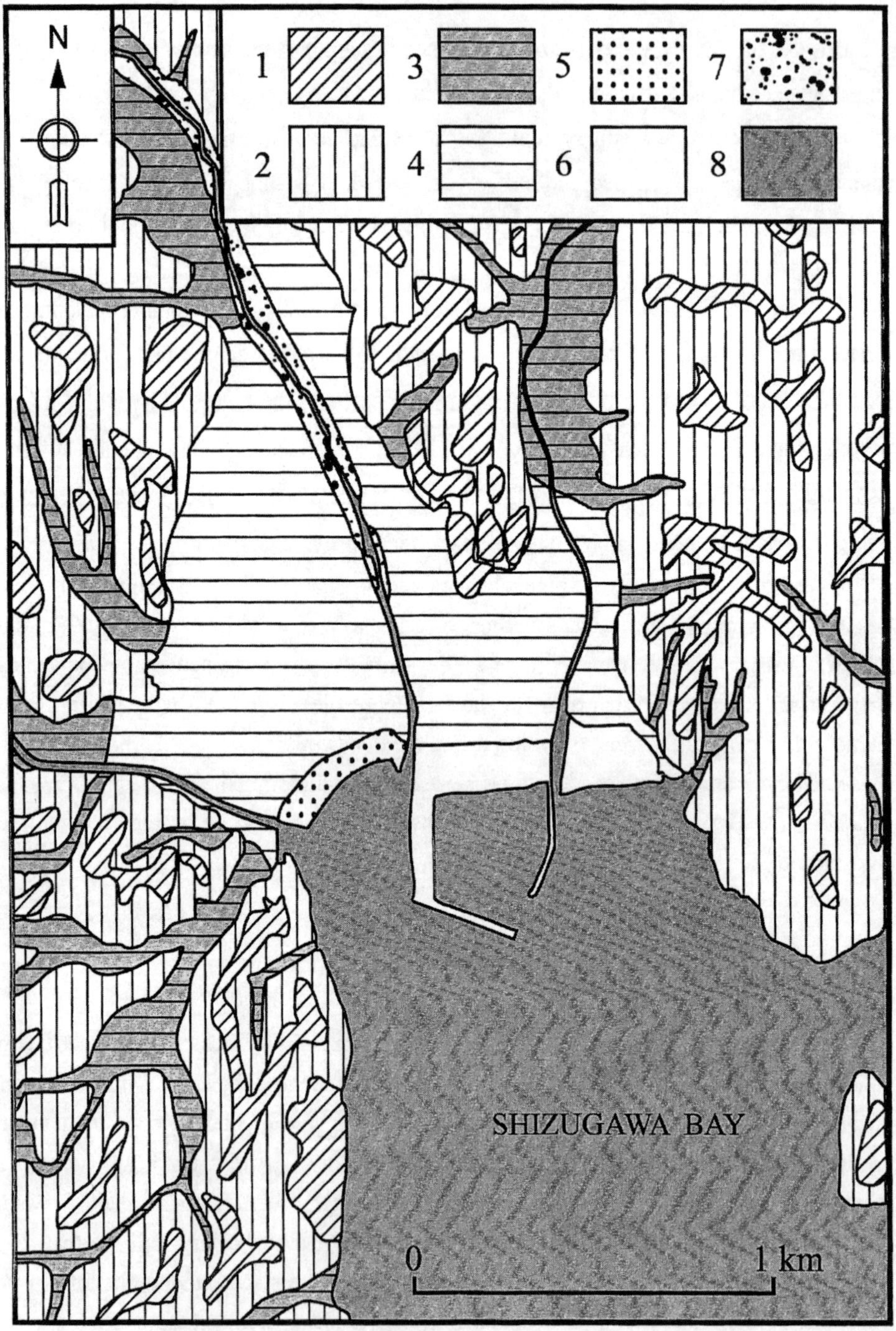

Figure 3.8 Geomorphological land classification map of the Shizugawa area (Oya, 1962; modified) (1) Terrace, (2) Steep slope, (3) Valley plain, (4) Delta, (5) Sand spit, (6) Reclaimed land or artificially filled field, (7) Dry riverbed, (8) River or sea.

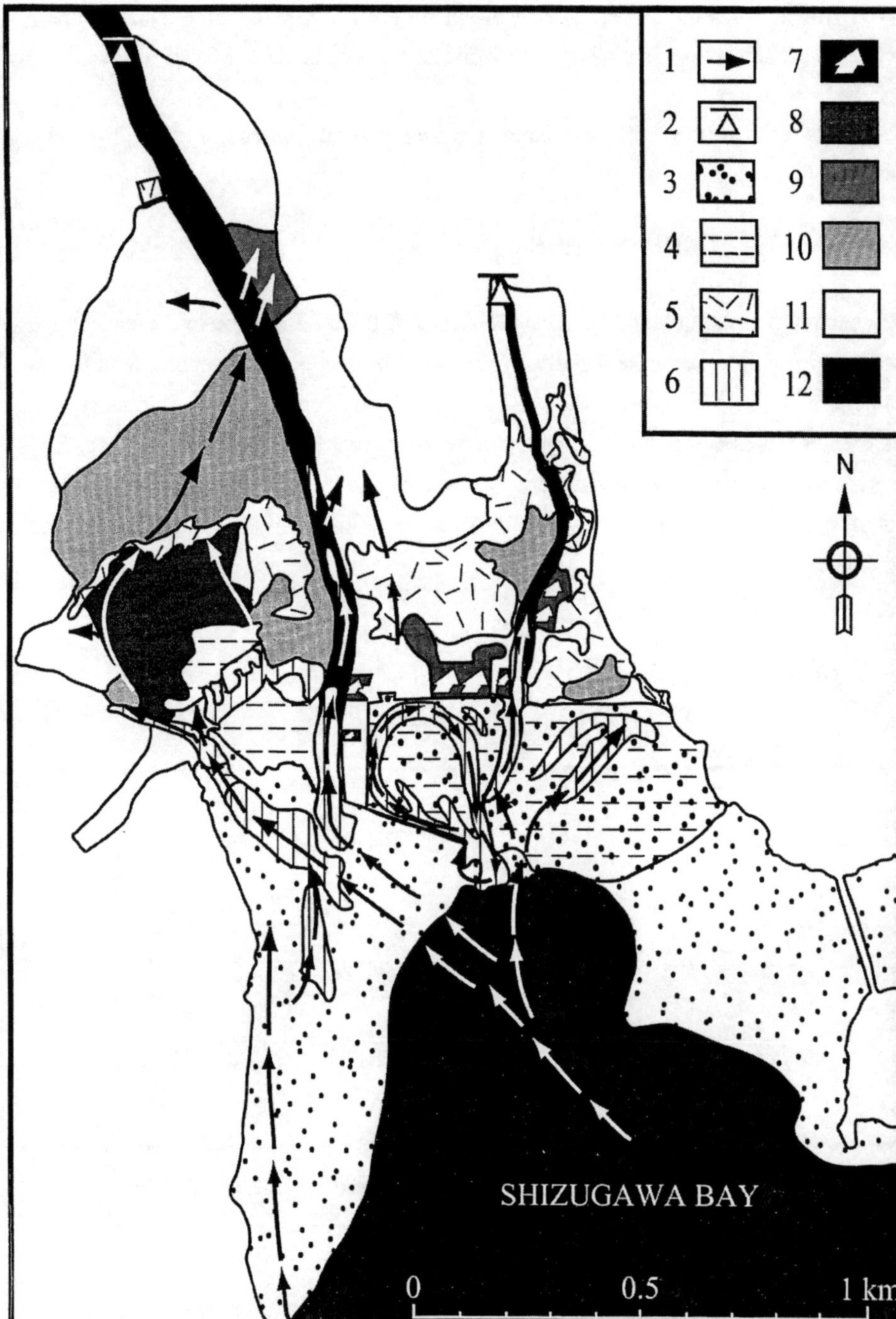

Figure 3.9 Map of the Shizugawa area portraying the effects of the1960 tsunami (Oya, 1962; modified) (1) Direction of tsunami, (2) Limit of tsunami intrusion, (3) Exposed seabed during period of tsunami, (4) Area of post-tsunami deposition, (5) Driftwood, (6) Area of erosion, (7) Area where homes were washed away, (8) Area submerged more than 60 days, (9) Area submerged more than 30 days, (10) Area submerged 4-5 days, (11) Area inundated less than 1 day, (12) River or sea.

Artificial waterways such as navigation routes can also become tsunami invasion routes. The nine-kilometer navigation route at Shiogama Port, Miyagi Prefecture, has served as such a path.

Tsunami flow through river courses, channels and sometimes through discharge ditches (*Figure 3.9*).

3.3.5. PROFILE OF INUNDATION

As tsunami proceed inland, the great resistance of the land surface to the current causes the water depth to decrease rapidly. A profile of how tsunami inundates land is shown in *Figure 3.10*. As each wave may last about 30 to 40 minutes, the profile must be considered as representing the maximum wave level at each point.

As the depth of water rapidly decreases, tsunami are eventually stopped by landforms. Generally speaking, their limits are established by the boundaries of valley and delta plains.

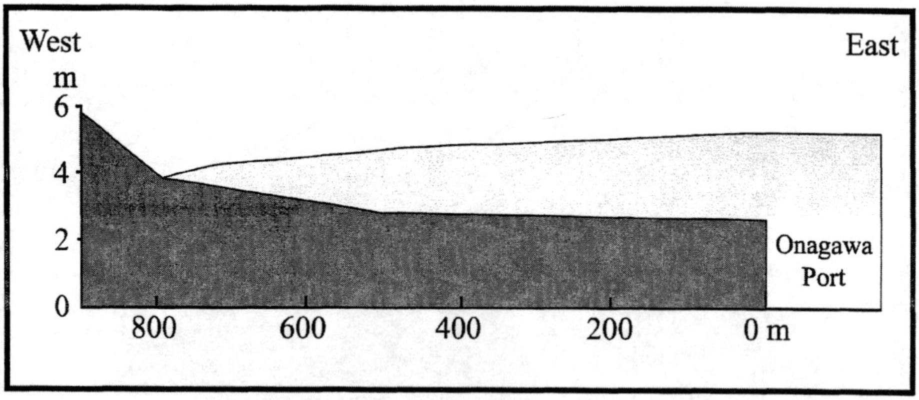

Figure 3.10 Profile of the tsunami at Onagawa (Oya, 1962; modified)

3.3.6. CONCLUSIONS

a) The Chilean tsunami is considered to be a model for tsunami research. The author analyzed its relation with embayments, submarine landforms and land, and can point out the submerged valleys, deep waterways such as mio, and rivers as the major routes of intrusion.

b) The mechanism of intrusion onto land resembles that of tidal waves.

c) The inundation on land is influenced by micro-landforms. If we analyze the physical properties of landforms beforehand, we can make predictions about disasters.

CHAPTER 4

RELATIONSHIPS BETWEEN GEOMORPHIC UNITS AND FLOOD TYPES OF RIVER BASINS

4.1. Geomorphic Features of Japan

Japan's most striking landform is its mountainous region, which includes some plains and intermontane depressions. The Hida, Kiso and Akaishi mountains, respectively called the Northern Japan Alps, Central Japan Alps and Southern Japan Alps, are found in central Honshu. These mountains are high and are surrounded by steep slopes (*Figure 4.1*).

Crustal movement in the Japanese archipelago has been extremely significant. Generally speaking, the mountainous region was formed by upheaval and the plains by depression. *Figure 7.2* shows vertical displacement during the Quaternary Period. At the end of the Tertiary Period the Japan Alps were only about 1,000 to 1,500 meters above sea level, but now their height is about 3,000 meters and the uplift continues.

In general, the rise in Japan has been about 1 or 2 millimeters per year, and numerous earthquakes have resulted from the frequent crustal movements. Some of the temblors have been devastating. For example, the Kanto Earthquake of 1923 destroyed the city of Tokyo and wreaked havoc throughout Kanagawa Prefecture and vicinity.

On the other hand, the plains regions have been lowered. For example, the Kanto Plain was depressed about 1,500 meters during the Quaternary Period, when rivers such as the Tone and Ara deposited sand and gravel in it, to a thickness of as much as 2,600 meters near Tokyo.

Since Japan is mountainous, various kinds of slope failures related to mass movement occur. When a slope collapses it can kill people and destroy the livelihood of the inhabitants of mountain villages.

In addition, there are more than 200 volcanoes in Japan.

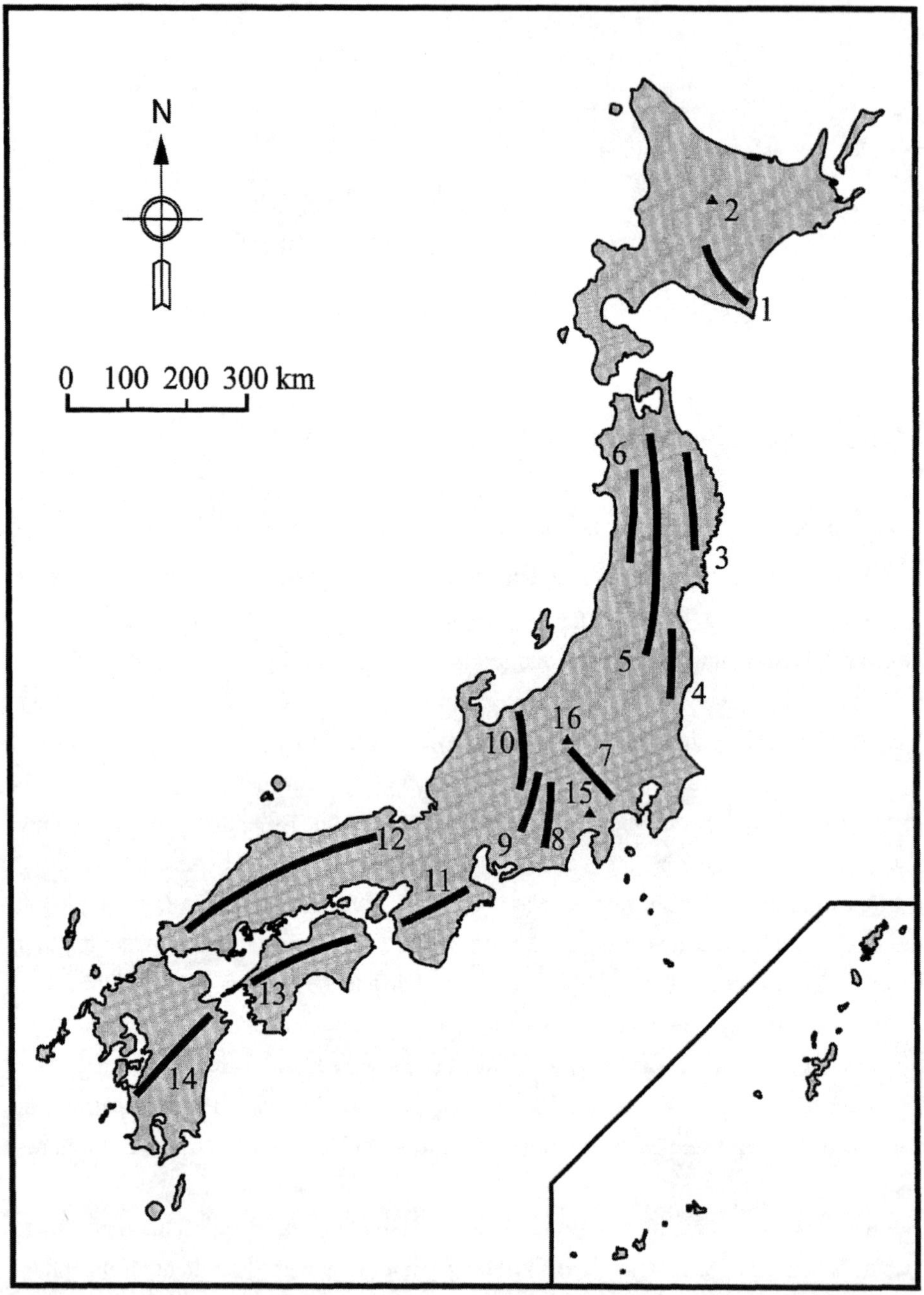

Figure 4.1 Topography of Japan
(1) Hidaka Mountains, (2) Taisetsu volcanic Mountains, (3) Kitakami Mountains, (4) Abukuma Mountains, (5) Ouu Mountains, (6) Dewa Mountains, (7) Kanto Mountains, (8) Akaishi Mountains (Southern Japan Alps), (9) Kiso Mountains (Central Japan Alps), (10) Hida Mountains (Northern Japan Alps), (11) Kii Mountains, (12) Chugoku Mountains, (13) Shikoku Range, (14) Kyusyu Mountains, (15) Fuji Volcano, (16) Asama Volcano.

4.2. Differences in Geomorphology Caused by Differing Climatic Conditions

Europe is located approximately between latitudes 35°N and 70°N, and according to the Köppen climatic division, it consists of Cs, Cfa, Cfb, Dfb, Dfc and BS. Japan lies between 24°N and 45°N, and consists of Cfa and Dfb (*Figure 4.2*).

Located within the northeastern part of the Monsoon Area, the Japanese archipelago experiences intense rainfalls and abundant precipitation. As a result, annual rainfall in Tokyo is 1,460 millimeters, Sapporo 1,460, Osaka 1,400 and Kagoshima 2,375 millimeters. In comparison, annual precipitation (averaged from 1959 to 1980) in Berlin is 590 millimeters, Paris 614, Moscow 657, London 759 and Vienna 914 millimeters.

While Europe has retained fossil topographies from the Tertiary and Pleistocene ages, in Japan rain and rivers have destroyed such features.

Further, Europe was glaciated at least four times during the Pleistocene Age by glaciers such as Günz, Mindel, Riss and Würm, and glaciers still remain locally in the Alps and Scandinavia. Japan lacks glaciers at present, and glaciated topography from the Pleistocene Age is limited.

Because of Japan's complex landforms and geology, rivers are short and their drainage basins extremely limited in area in comparison with those of continental areas. There is no river longer than 400 kilometers. The country's largest river basin is that of the Tone River near Tokyo, which has a length of 322 kilometers and a basin area of about 15,760 square kilometers. The short lengths of the rivers and the sharp elevation differences in the mountains tend to produce steep river gradients, even near the river mouths at the coast.

As a result of these geomorphological features, river flow varies rapidly in response to fluctuations in rainfall and snowmelt. In particular, an extremely rapid increase in discharge from heavy rainfall can cause hazards such as inundation, slope failure and debris flow. Typically, a river about 100 kilometers long will have a time lag between rainfall and peak discharge, as recorded by hydrograph, of only about six to eight hours.

Due to its numerous typhoons, the coastal areas of Japan are frequently affected by tidal waves as well as by tsunami from earthquakes. These high waves can cause severe damage as almost all the large cities and industrial regions of Japan are found along bays. As well, human activity has contributed to a significant amount of coastal erosion.

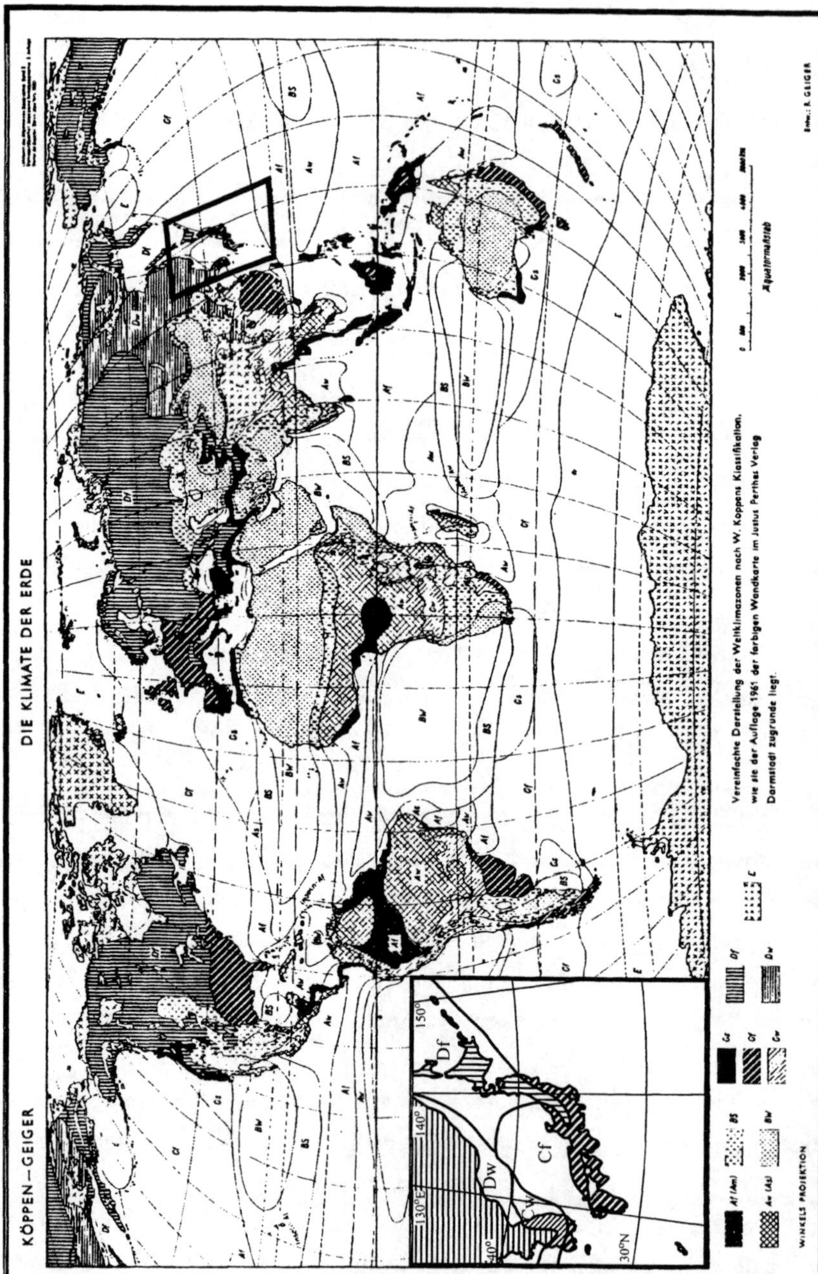

Figure 4.2 World's climatic zones according to Köppen classification (Geiger, 1965; modified)

4.3. Features of Japan's Alluvial Plains

4.3.1. BASIC FORM OF ALLUVIAL PLAINS

The normal method of formation of the fluvial plains is considered to be as follows: First, a natural levee forms along a river course; second, new channels form, partly because of variations in river flow and partly from changes in the volume of sand and gravel transported by the river. Next, natural levees form along the new river course.

The natural levees consist of sand and gravel in the upper parts of the plains and of sand and silt or clay in the lower. As the coefficient of adhesion of sand and gravel is less than that of sand, silt or clay, the levees of the upper parts of the plain are readily destroyed by the streams. Also, many streams will be formed in the upper parts of the plains, typically with braided courses. Finally, the sand and gravel in the upper parts of the plains form alluvial fans (Oya, 1973a, b).

From the lower edges of the fans to the lower segments of the streams, the natural levees consist of sand. As the coefficient of adhesion of sand is greater than that of sand and gravel, there is less bank erosion in the lower regions of the plains than in the upper. In those lower regions, the stream courses are more stable and the streams meander, with well-developed natural levees. Back-marshes are found between the levees. They consist of silt or clay.

Near the seacoast, the natural levees are smaller because of marine action, i.e. the ebb and flow of daily tides. These areas, which consist of silt and clay, can be considered as deltas. Toward the seacoast, the streams straighten and cut deep trenches in the plains. While of course there are local differences, the standard form of the fluvial plains along the coast is as follows:

Fan + Natural Levee + Delta
(Back-marsh)

4.3.2. INFLUENCE OF EUSTATIC MOVEMENT AND HUMAN ACTIVITY

Generally speaking, the sea level was about 100 to 140 meters lower about 18,000 years B.P. Then, about 6,000 to 4,000 years B.P., it rose to about 4 to 6 meters above its present level. In Japan, this rise is called the Jomon Transgression. Next, about 2,000 to 2,100 years B.P., the sea level dropped to about 3 meters lower than it is now. After that the sea again rose, with minor fluctuations, to its present level (*appended Table 3*).

Elevations of the Japanese plains are low, especially near the deltas. For example, in the Nobi Plain, the distal edge of the Kiso River fan is 12 meters above sea level and the natural levee between 1 and 12 meters, whereas the delta is 1 meter above to 2 meters below sea level—an indication of the remarkable influence of the Jomon Transgression upon the deltas.

Rice has been the base of the Japanese diet from prehistoric times till present. In the Jomon Age, production was small, but it increased during the Yayoi Period. The back-marshes were utilized for paddy fields, and the people lived on the natural levee areas partly because they were free from flooding and partly because of the convenient navigation available there. At that time deformation of the plain was but minor.

Following the Edo Period, a great deal of reclamation of land from the shallow sea took place, especially around Tokyo Bay, Ise Bay, Osaka Bay, Ariake Sea, etc. For example, at the mouth of the Kiso River in the northern part of Ise Bay, the area of reclaimed land expanded quickly; the coastline advanced an average of 44 meters per year.

Since the Meiji Restoration, much construction has taken place on landfill along the bay facing the big cities, including Tokyo, Osaka and Nagoya. Such land is used largely for industry and wharves. This expansion of industrial regions has caused the ground surface in the deltaic area to lower, chiefly from the overpumping of ground water. Due to this, the area below sea level has expanded around Tokyo, Nagoya and Osaka. To stop or slow this lowering, the use of ground water for industrial purposes was prohibited, and factories were requested to use the municipalities' water supplies instead. As a result, the use of ground water has almost ceased.

4.4. Overflowing Type: The Kiso River

The Kiso River rises in the southeastern part of the Hida Mountains and is joined by the Otaki River, which originates on the Ontake Volcano. The Kiso then flows between the western edge of the Kiso Mountains and the eastern edge of the Atera Mountains. After being joined by the Hida River, it crosses the Nobi Plain, where it has formed large alluvial fans, natural levees and deltas. It then continues to Ise Bay (Oya, 1956).

This triple combination of geomorphic units

Fan + Natural Levee + Delta
(Back-marsh)

indicates that river deposition is remarkable.

A cross section of the Kiso River and its surrounding plain shows that the land becomes higher as one nears the river (*Figure 4.3*). Therefore, when a flood occurs the water spreads beyond the natural levee to the back-marsh or delta. There has been no severe flooding in the last 30 years, but during the Ise Bay Typhoon the water flowed upstream through the Kiso, Nagara and Ibi rivers, breaking through the embankments and overflowing onto the adjacent lowland (Oya, 1973a, 1988a).

There is a close relationship between the topographic features of alluvial plains and mountains. About 53.5 percent of the drainage area of the Kiso River is above 700 meters, and more than 52.2 percent of that drainage area is steeper than 30 degrees. The relief energy of the area is high, with a maximum of 1,300 meters (*Figure 7.3*). During the Quaternary Period, the mountains were raised about 2,000 meters, i.e. 60 to 70 percent of their present height. Uplift has continued at a rate of 5 to 11 centimeters in the last 50 to 60 years. The principal rocks in the mountains include granite and andesite, which weather readily.

The geomorphology of the Kiso River basin results in the provision of a large and continuous supply of sand and gravel. Further, these features cause landslides during severe rainfall, which also deposit a large quantity of debris on the river bottom. Even the reservoirs for hydroelectric power stations at Yaotsu and Oi face a great problem of silting owing to the heavy influx of sand and gravel from the river. Otherwise, as the river runs along a continuous canyon without any intermontane depressions, the sand and gravel from the landslides are carried directly to the plain.

A large alluvial fan, with its apex at the city of Inuyama, was formed in this plain. Several natural levees lie south of the fan margin. When a riverbed is greatly raised, the shifting of the river brings large floods that form new natural levees along its channel.

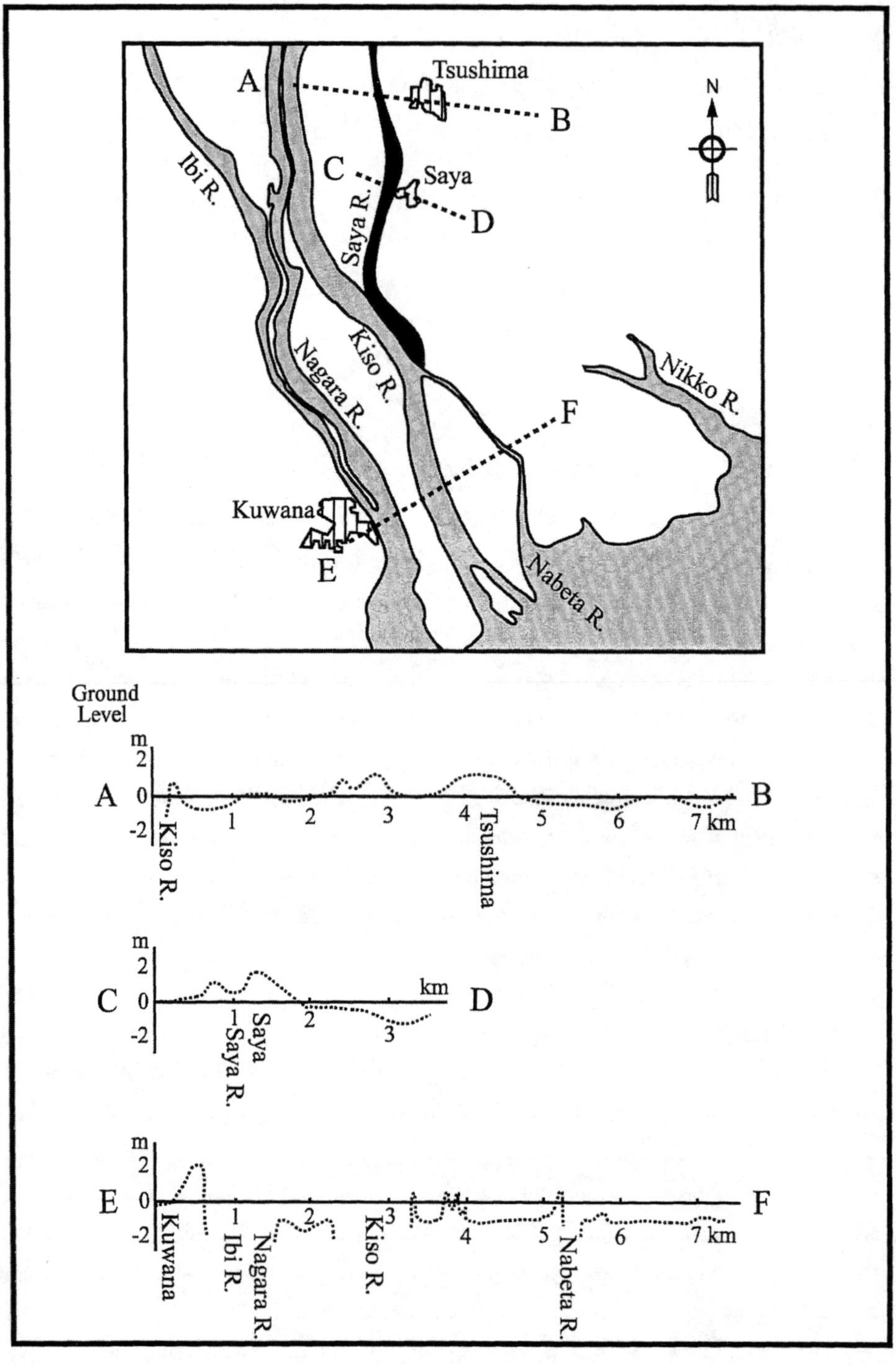

Figure 4.3 Topographical cross section of the lower reaches of the Kiso River basin (Oya, 1973a; modified)

In the Nobi Plain, several divergent natural levees formed by this process can be easily recognized (*Figure 3.1*).

Although the Nobi Plain is susceptible to flooding, its topography is beneficial in the utilization of the river for irrigation and as a water supply throughout the plain.

A similar overflowing pattern is seen in the Han River, which runs through Seoul, and the Nakdong River in South Korea.

4.5. Concentration Type (Retarding Type): The Chikugo River

The Chikugo River—the largest river in Kyushu—originates on the northern flank of Aso Volcano and flows down to the Ariake Sea. From the river's origin to the Yoake Canyon the region is mountainous with several intermontane depressions, and from the canyon to the river mouth the region is a plain. In the plain, the Chikugo River has formed a small fan, and tributaries from the north and south have formed larger ones. Even natural levees, which are common below the fan, are rare here. The middle and lower delta, which is developed along the main course, is only about seven kilometers wide (*Figure 4.4*). The combination of geomorphic units is as follows:

Fan
↓
Tiny Fan + Delta + Tidal Delta + Subaquatic Delta
↑
Fan

If we compare these geomorphological features with those of the Kiso River basin, we can see distinct regional differences. For example, the mountains in the Kiso River basin are higher than in the Chikugo basin. Similarly, the slope of the Chikugo River basin, as might be expected, is less than that of the Kiso. The maximum relief energy is a mere 800 meters in the former, but 1,300 meters in the latter. Hence the Chikugo River basin is less susceptible to landslides than that of the Kiso.

There are several intermontane depressions, including the Oguni and Hida, along the main course of the Chikugo River, and the Handa Plateau and Kusu Depression along the Kusu River, which is a tributary to it. When a river reaches a depression, much of the

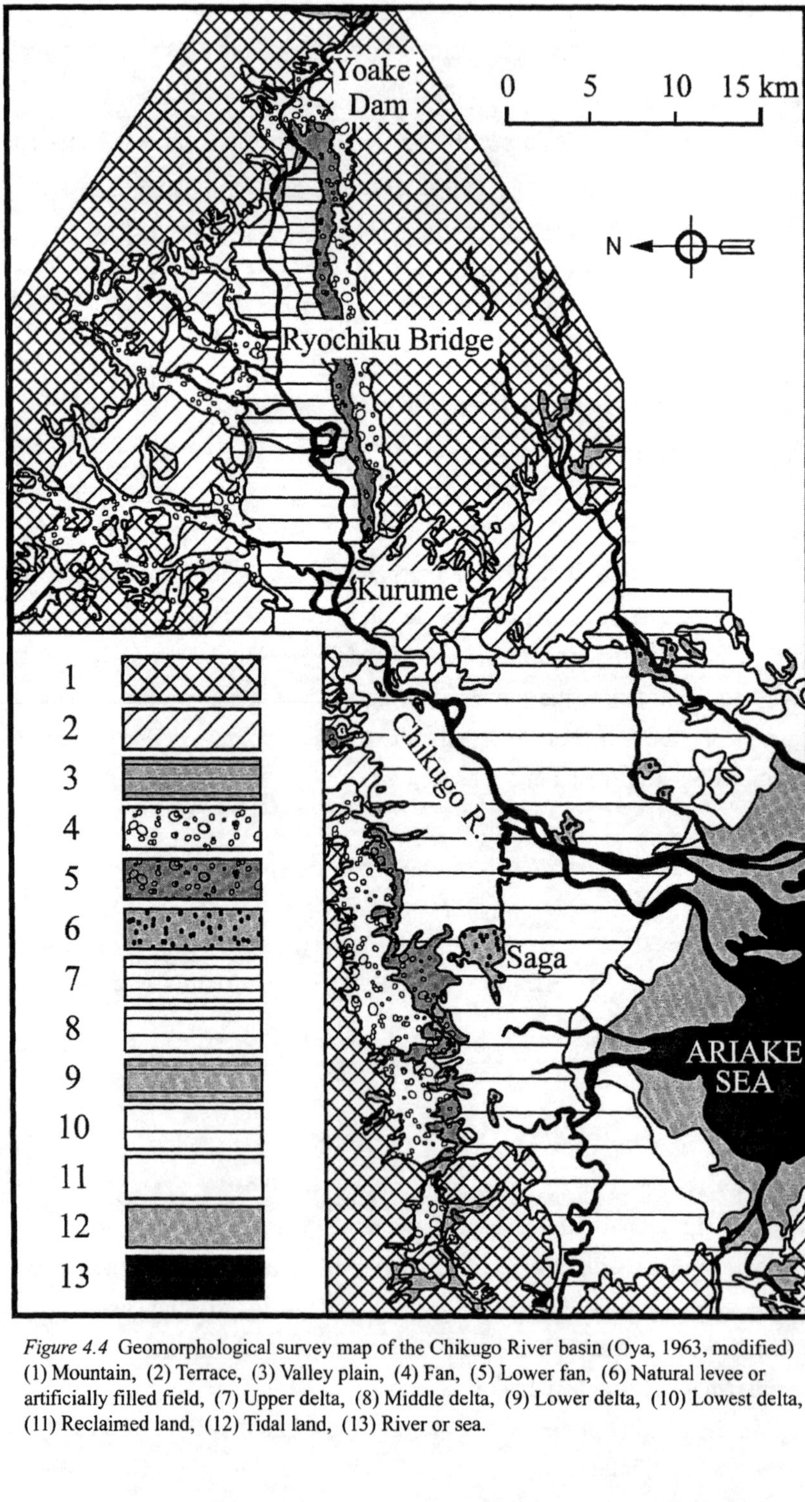

Figure 4.4 Geomorphological survey map of the Chikugo River basin (Oya, 1963, modified) (1) Mountain, (2) Terrace, (3) Valley plain, (4) Fan, (5) Lower fan, (6) Natural levee or artificially filled field, (7) Upper delta, (8) Middle delta, (9) Lower delta, (10) Lowest delta, (11) Reclaimed land, (12) Tidal land, (13) River or sea.

coarser gravel transported from its upper reaches is dropped in the depression, and not much is carried further downstream. Leveling from 1893 to 1963 shows that most of the upper part of the Kusu Depression sank 9.19 centimeters during those 70 years, whereas the canyon located in the lower reaches of the depression rose 9.2 centimeters.

This crustal movement has been partly responsible for stopping sand and gravel from being carried to the river's lower reaches. A study of the silting of the reservoirs of the Yoake Dam shows that the deposition rate there is only 1/42 of that of the Kiso River (Tada et al., 1957).

In the middle and lower reaches of the Chikugo River, the land lowers toward the river (*Figure 4.5*). This is mainly due to erosion, to dips in the ground along the river course and to the much greater deposition by the tributaries along the foothills than nearer the main river course. Owing to this peculiar topography, floodwater flows from the higher surrounding area into the river. In fact, when overflow does occur, the floodwater does not disperse but flows back into the river farther down. This type of flooding was observed during the flood of 1953, when the left bank dikes broke at Eri and floodwater invaded the protected area, subsequently breaking through the left dike from its back side to rejoin the main river.

The floods of 1921, 1935 and 1941 caused shortcuts in the meandering section of the Chikugo River, which resulted in the shortening and straightening of the river. Because of this and the building of continuous embankments, river velocity increased. Whereas between 1925 and 1932 it took the river 9.2 hours to flow from Hida to Kurume, from 1942 to 1952 it took only 4.8 hours. This increased velocity of the flood current lowered the riverbed between the river mouth and Ryochiku Bridge, which is 45 kilometers upstream from the river mouth, by 43 centimeters from 1923 to 1961. This lowering was abetted by both man-made construction, which affected the velocity of the river, and by excavation of sand and gravel from the riverbed. Thus, in the middle reaches of the Chikugo River, erosion is more important than deposition due to changes brought on by human work.

In this area inundation is minor and deposition limited. However, the topography makes it inconvenient for water use. Moats, which are well developed in the lower reaches of the Chikugo River, are countermeasures to this condition of water shortage. Flood types similar to those of the Chikugo River occur in the Shonai and Yamagata intermontane depressions from the Mogami River and in the Niigata Plain from the Agano (Tada et al., 1957; Oya, 1973b, 1977a).

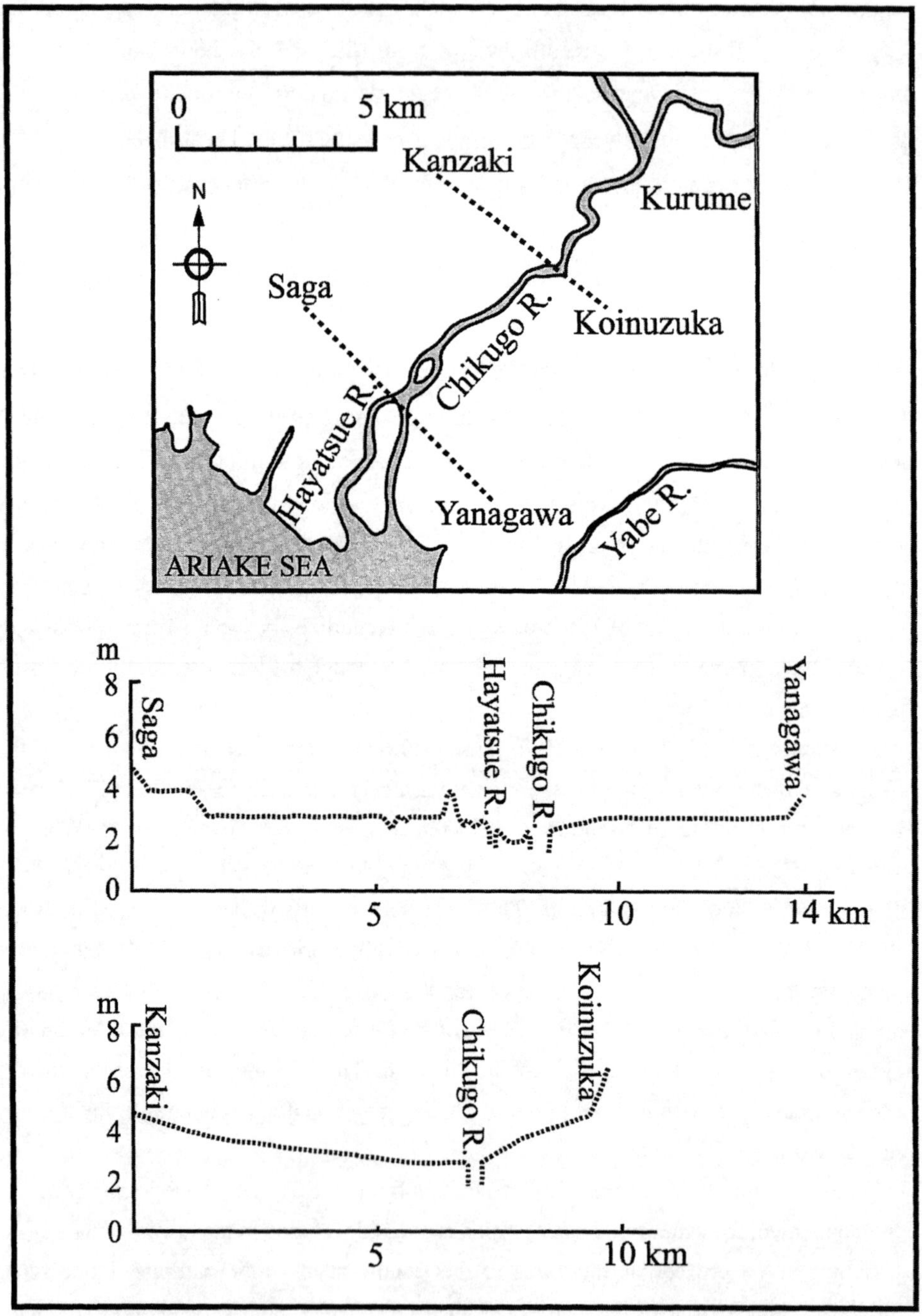

Figure 4.5 Topographical cross section of the lower reaches of the Chikugo River basin (Oya, 1973a; modified)

4.6. Combination of Overflowing and Concentration Types: The Kano River

A third type, which is rather rare, can be created by combining the two main types.

The upper and lower parts of the Kano River Plain in the Izu Peninsula are separated by narrows in the river channel. Those narrows are formed by Mount Kanuki and by the Mishima Fan, which consists of mudflow from Fuji Volcano.

Natural levees and back-marshes are developed in the Takata Plain. This area was covered by seawater in an earlier alluvial epoch (around 6,000 years B.P.). At that time, thick marine clay was deposited. Afterward, regression by the sea, eject from Fuji Volcano and fluvial deposits of the Kano River caused this area to become land. The natural levee area becomes submerged during extreme flooding, but the water is shallow and drains off well. In contrast, when the back-marsh and deltaic areas are flooded, the floodwater is deep and its stagnation is long.

The Takata Plain looks like a depression because its elevation in its lower reaches, i.e. Numazu Narrows at about 12 meters above sea level, is higher than in its upper reaches. The lowest part of the depression, at 7 meters, is the Nagasaki Village area. When there is a flood, that area is inundated (*Figure 4.6*).

The Numazu Coastal Plain is a delta with coastal sand dunes. There are no natural levees.

From Sept. 26 to 27, 1959, the Kano River Typhoon hit the Izu Peninsula, passed the northern part of Tokyo and then proceeded northeast along the coast of the Pacific Ocean. In the town of Yugashima, in the central part of the peninsula, rainfall reached 750 millimeters between 09:00 on the 25th and 24:00 on the 26th. This caused a flood of 3,000 to 5,000 cubic meters per second at Ohito Bridge, on the lower reach of the Kano River. As a result, 671 people drowned, 182 were reported as missing, 254 homes were damaged, and 625 homes and 76 bridges were washed away.

Distinct regional differences in the type of inundation caused by the different geomorphologies could be seen (*Figure 4.7*). In the Takata Plain, floodwater from the main river overflowed the natural levees and flowed down to the back-marsh or delta regions. In contrast, in the Numazu Coastal Plain there was no inundation caused by river flooding. High waves advancing from the mouth overflowed part of a delta, but the water soon rejoined the main river (Oya, 1960).

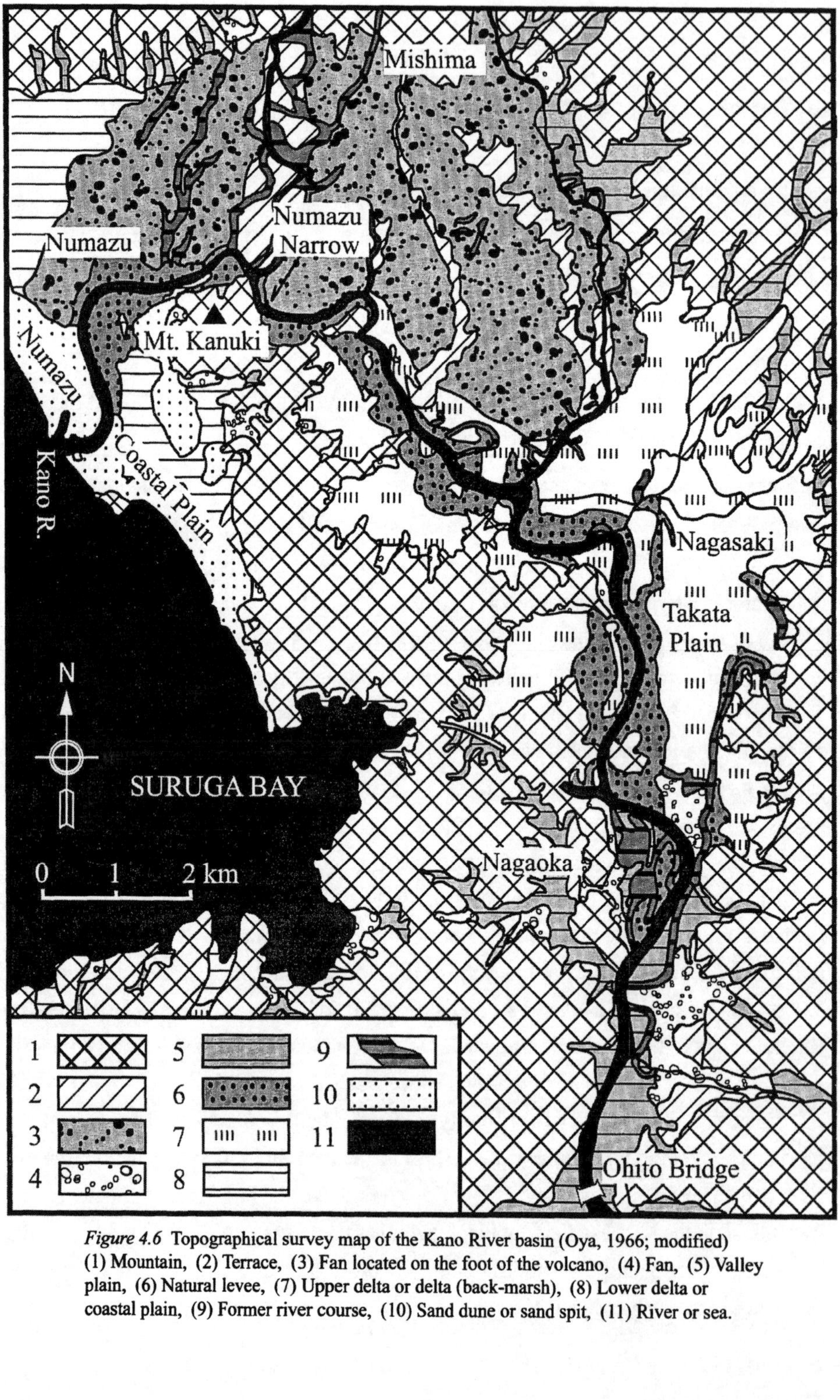

Figure 4.6 Topographical survey map of the Kano River basin (Oya, 1966; modified) (1) Mountain, (2) Terrace, (3) Fan located on the foot of the volcano, (4) Fan, (5) Valley plain, (6) Natural levee, (7) Upper delta or delta (back-marsh), (8) Lower delta or coastal plain, (9) Former river course, (10) Sand dune or sand spit, (11) River or sea.

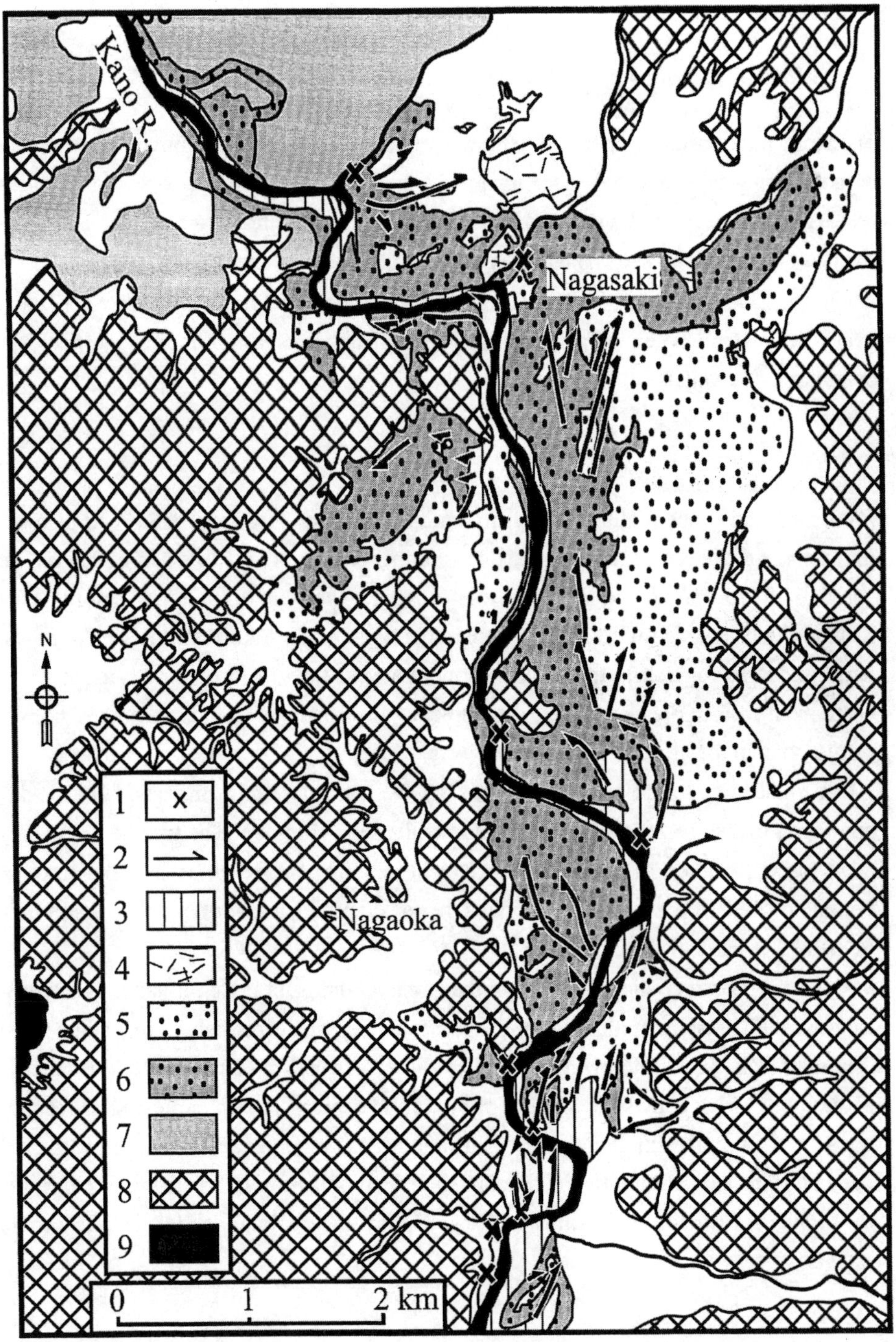

Figure 4.7 Map of the Kano River basin in the Izu Peninsula showing the flooding that occurred on Sept. 26, 1958 (Oya, 1966; modified)
(1) Destroyed embankment, (2) Flood current direction, (3) Erosion or deposition, (4) Driftwood, (5) Thin layer of silt and loam deposition, (6) Thick layer of silt and loam deposition, (7) Inundation caused by rainfall in protected areas, (8) Mountain or terrace, (9) River or sea.

4.7. Natural Conditions in the Korean Peninsula

The Korean Peninsula is an Epeirogenic Movement Area, so crustal movement was of a broad warping type and took place long ago. Therefore, the drainage basins of its rivers are larger and the rivers longer than in the Orogenic Movement Area of Japan. Three big rivers—the Han, Kum and Nakdong—flow south, and a considerable part of South Korea lies in their basins (*Figure 4.8*). Because of large-scale deposition, all three rivers have overflowing flood types in each plain. This deposition is due to climatic and geologic conditions, and vegetation features.

4.7.1. CLIMATIC CONDITIONS

Figure 4.9 shows the number of days in a mouth in which the temperature was below 0 for an entire day, i.e. "frozen days," and the number of days below 0 for part of the day, i.e. "partially frozen days." The graph shows the data for the year 1940 in two regions of the Korean Peninsula: Shinuiju in the northernmost part and Mokppo in the southernmost part. Temperatures during the winter are more severe than in Japan, and "frozen days" occur in all parts of the peninsula. The number of "frozen days" in mountainous regions is greater than in the cities.

Due to the large temperature differences between summer and winter and between day and night, rocks weather strongly and land collapse is common. The weathering mostly takes place in winter, as water enters cracks in the rock and freezes, which increases its volume by 4 to 10 percent. This fractures the rocks, producing a lot of debris, especially in the higher parts of the mountains.

The main characteristic of rainfall in the Korean Peninsula is that it is not evenly distributed, with most rain falling from June to September, and especially in July and August. The rainfall is light in the winter months. These seasonal differences are illustrated in *Figure 4.9*, which shows the amount of rainfall in a given month expressed as the percentage of the total rainfall in the year. July and August together account for more than 51 percent of the annual rainfall and the four summer months for 71.3 percent, as compared to 9.6 percent for the four winter months. The average annual rainfall is 1,269 millimeters.

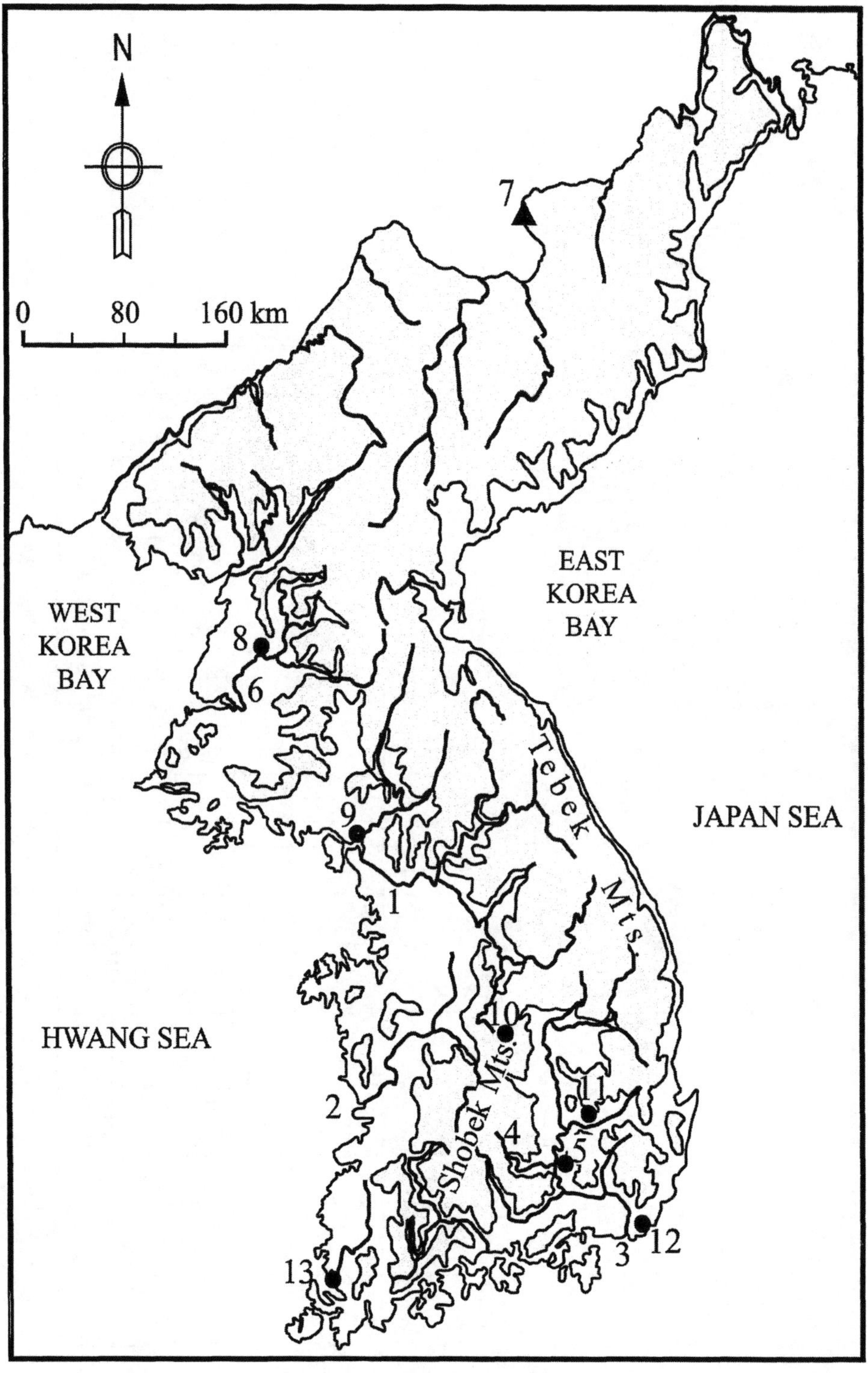

Figure 4.8 Topography of the Korean Peninsula (Oya, 1971; modified)
(1) Han River, (2) Kum River, (3) Nakdong River, (4) Hwang River, (5) Uppo Lake, (6) Dedong River, (7) Mount Bekudo, (8) Pyongyang, (9) Seoul, (10) Chupungnyang, (11) Taegu, (12) Pusan, (13) Mokppo.

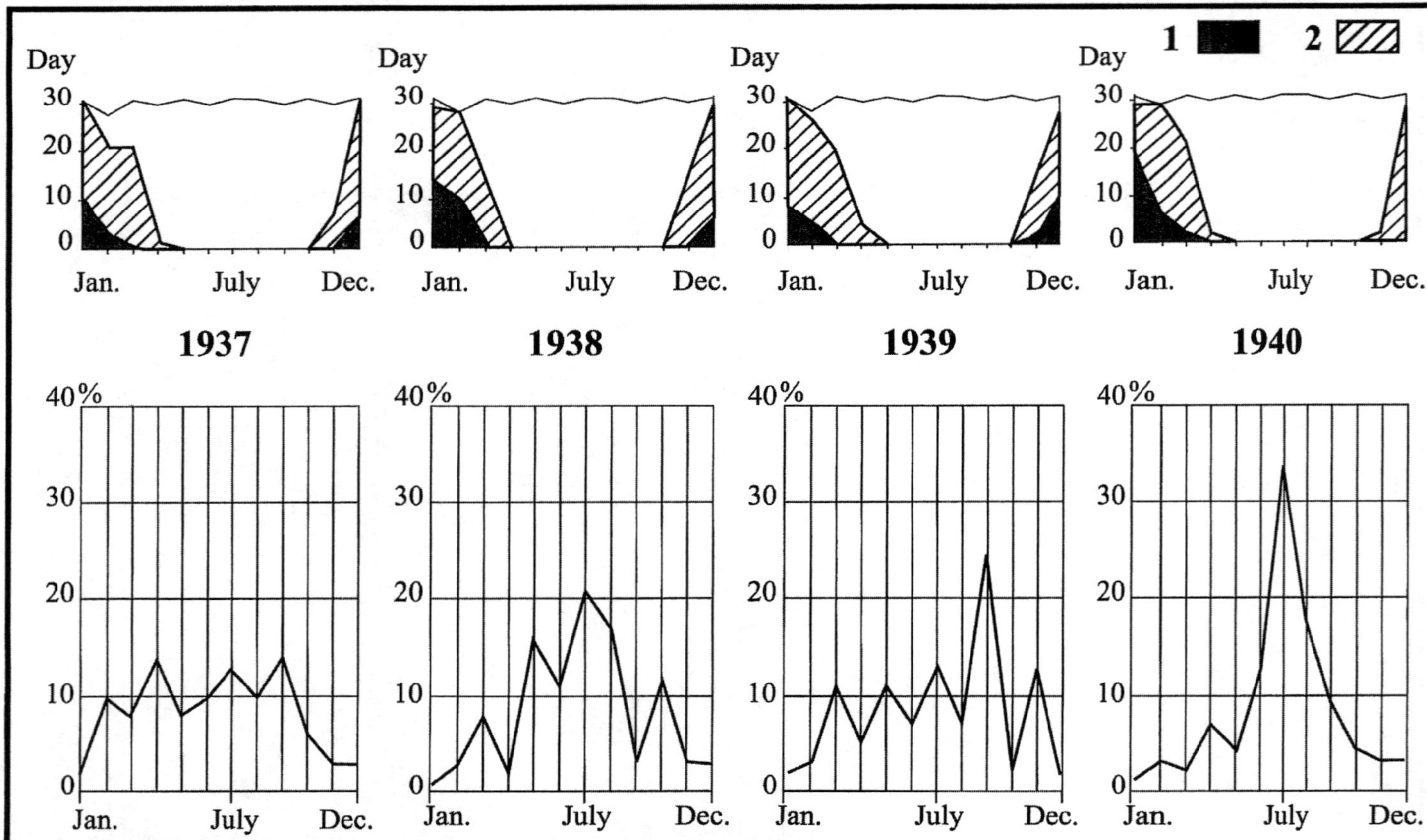

Figure 4.9 Number of "frozen days" and "partially frozen days" each month, and ratio of monthly rainfall to annual at Chupungnyang (Oya, 1971; modified) (1) Frozen days, (2) Partially frozen days.

4.7.2. GEOLOGIC CONDITIONS

There are many igneous rocks, especially granite, distributed in the Tebek Mountains and Shobek Mountains. Such rock types are highly susceptible to mechanical weathering. Sandstone and black shale beds, found in the southeastern part of South Korea, i.e. the Nakdong River basin, also have many cracks and so are also susceptible to mechanical weathering.

The status of forestry in the peninsula's mountains is generally poor because of physical conditions and deforestation. As stated above, mechanical weathering of the rock in the mountainous region is severe in winter, and debris is washed away by the torrential summer rainfall. Thus, little soil is present. It is clear that forestation in such climatic and geological zones is difficult, particularly after deforestation. About 2,000 years ago, forestry was well developed.

4.8. Overflowing type: The Nakdong River

The Nakdong is the largest river in South Korea. The combination of the geomorphic elements in its plain is as follows:

Pediment + Peri-pediment + Natural Levee + Delta + Tidal Land
(Back-marsh)

Alluvial fans are common in the plains of Japan, but rare in the plains of the Korean Peninsula, including the Nakdong River basin.

There are many pediments or pediplains at the foot of the mountains in the Nakdong River basin. These pediments are covered by block streams, formed by mechanical weathering in the winter and torrential rainfall in the summer.

Natural levees, however, are well developed. *Figure 4.10* shows the geomorphological survey map of the lower reaches of the Nakdong River (Uppo Lake and vicinity). The Uppo Lake and back-marshes occupy the spaces behind the natural levees. In former days, floodwater from the main river course would overflow the natural levees into the back-marsh area, which would remain submerged for a long time. Now, sluice gates

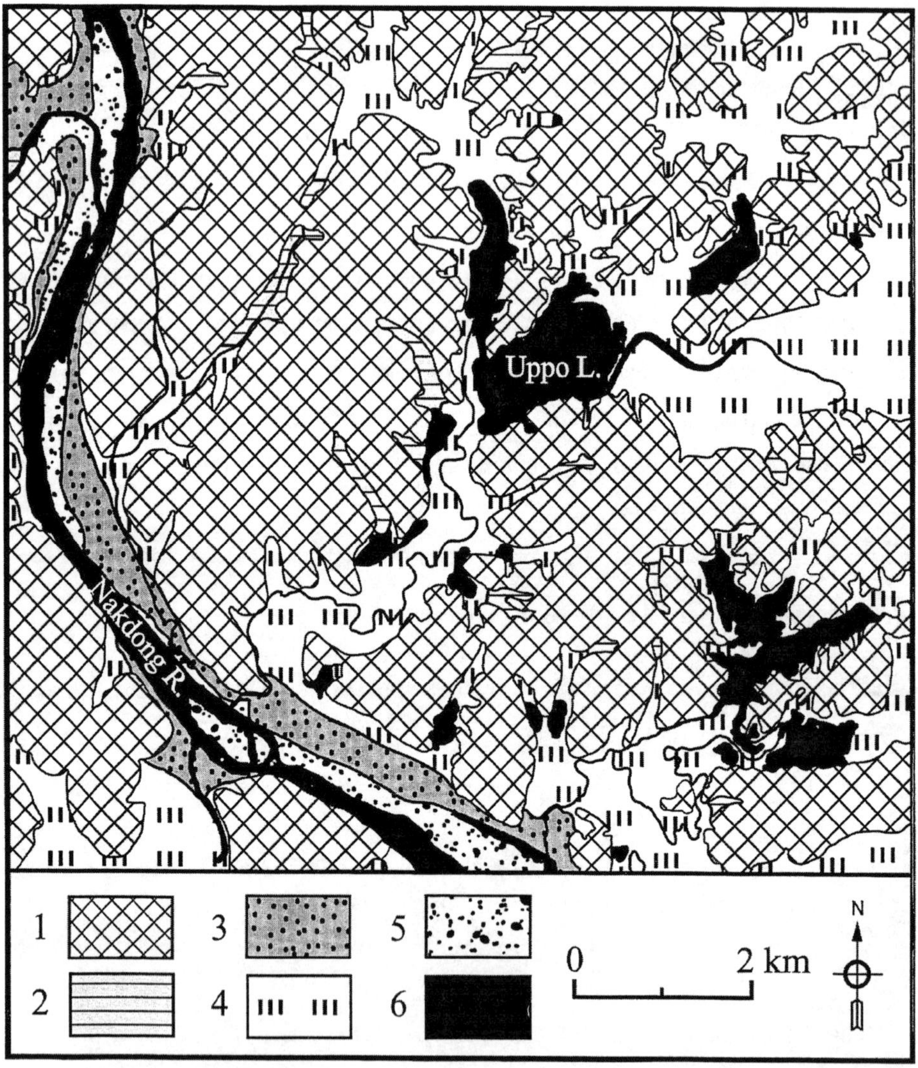

Figure 4.10 Provisional geomorphological survey map of the lower reaches of the Nakdong River, South Korea (Oya, 1971; modified)
(1) Mountain, (2) Valley plain, (3) Natural levee, (4) Delta or back-marsh, (5) Dry riverbed, (6) River or lake.

constructed at the junctions are closed to stop the floodwaters from advancing into the tributary, and opened after the water level in the main river course has fallen.

Deposition by a flood in 1968 showed how natural levees and back-marshes form. That flood raised the height of the natural levee near Koryon Bridge by about 1.5 meters, but only about 30 centimeters of deposits were laid down in the back-marshes. During a flood, the inundation is relatively shallow over the natural levees, but deposition is large. Therefore, almost all the big cities and towns are located on the pediment or piedmont gentle slopes.

The Hwang River has the greatest deposition of all the tributaries of the Nakdong River. An area of reclaimed land surrounded by an artificial levee is found in the upper part of Bak-Sil Pond, near the town of Hyopchun. For most of the year, this reclaimed land is covered by water, and only in summer does it become a paddy, through the use of pumps. Prior to 40 years ago there was no need to surround the area by levees. But due to the increasing elevation of the mainstream, the water level of the pond became higher and the construction of levees became a necessity.

Jungyang Pond now exists across the river from Hyopchun. It was formed 30 years ago, when the raising of the natural levees along the main course dammed small tributaries. According to the people living in the area, the riverbed has risen about 1.5 meters during the past several decades. The area was devastated by a flood in 1956 that caused much deposition along the river. An especially large deposit of about 2 meters was seen at the town of Munriniri along the river.

In summary, deposition is strong in the lower reaches of the Hwang River. When flooding occurs, much deposition takes place and floodwater overflows from the main course to the adjacent lowland.

4.9. Studies of Overflowing-Type Flooding from the Han River

There are important natural levees along the Han River, which runs through Seoul, whereas only back-marshes lie along its tributaries. This shows that the deposition along the main river course is much greater than in the tributaries. In addition, floodwater overflows from the main river and into the tributaries.

Park No Shik has studied the geomorphological land classification of the Han River (*Figure 4.11*) and Kim Tae Hi its stages of inundation (*Figure 4.12*). After a great flood devastated the area in July 1972, Kim prepared maps showing the state of flooding and the stages of inundation. A comparison of the geomorphological land classification map

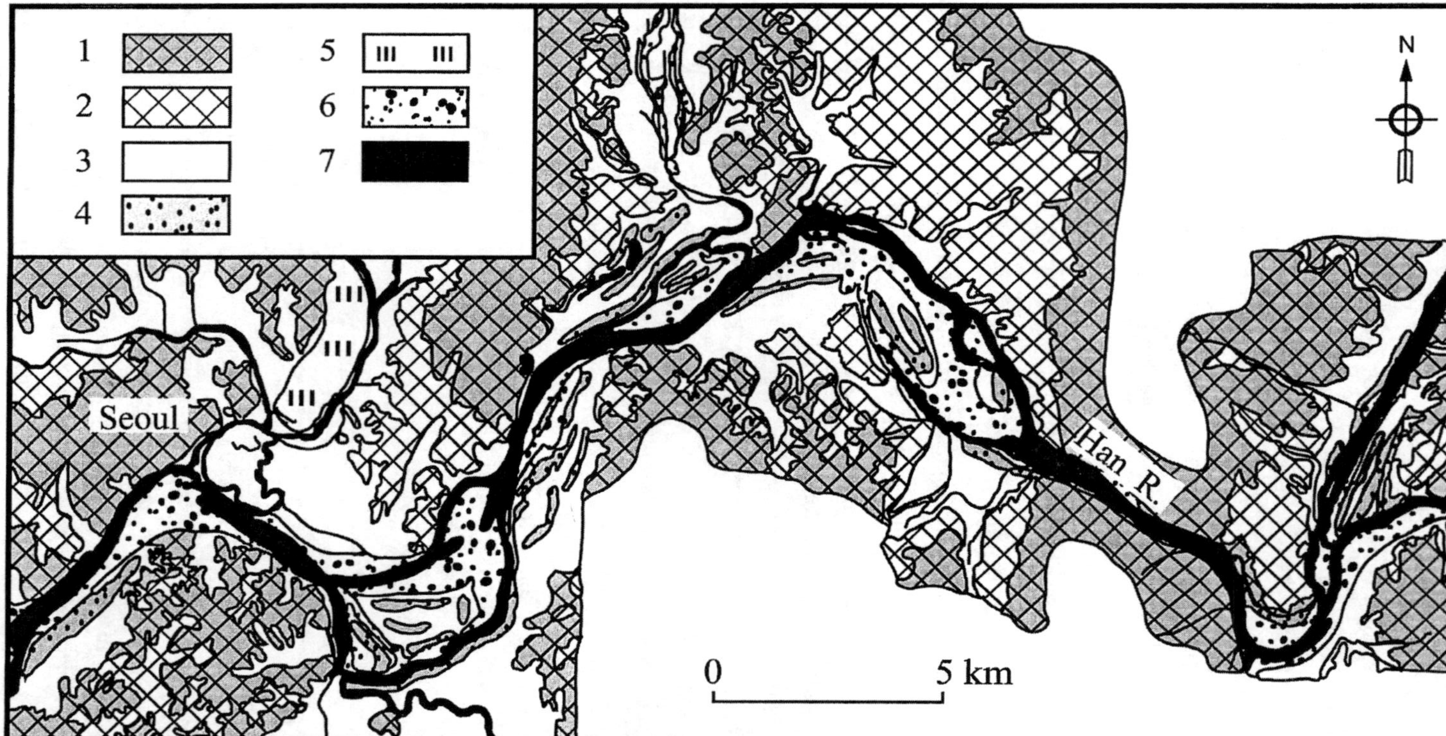

Figure 4.11 Geomorphological land classification map of the lower reaches of the Han River, South Korea (Park, 1967; modified)
(1) Mountain, (2) Pediment, (3) Alluvial plain, (4) Natural levee, (5) Marsh, (6) Dry riverbed, (7) River

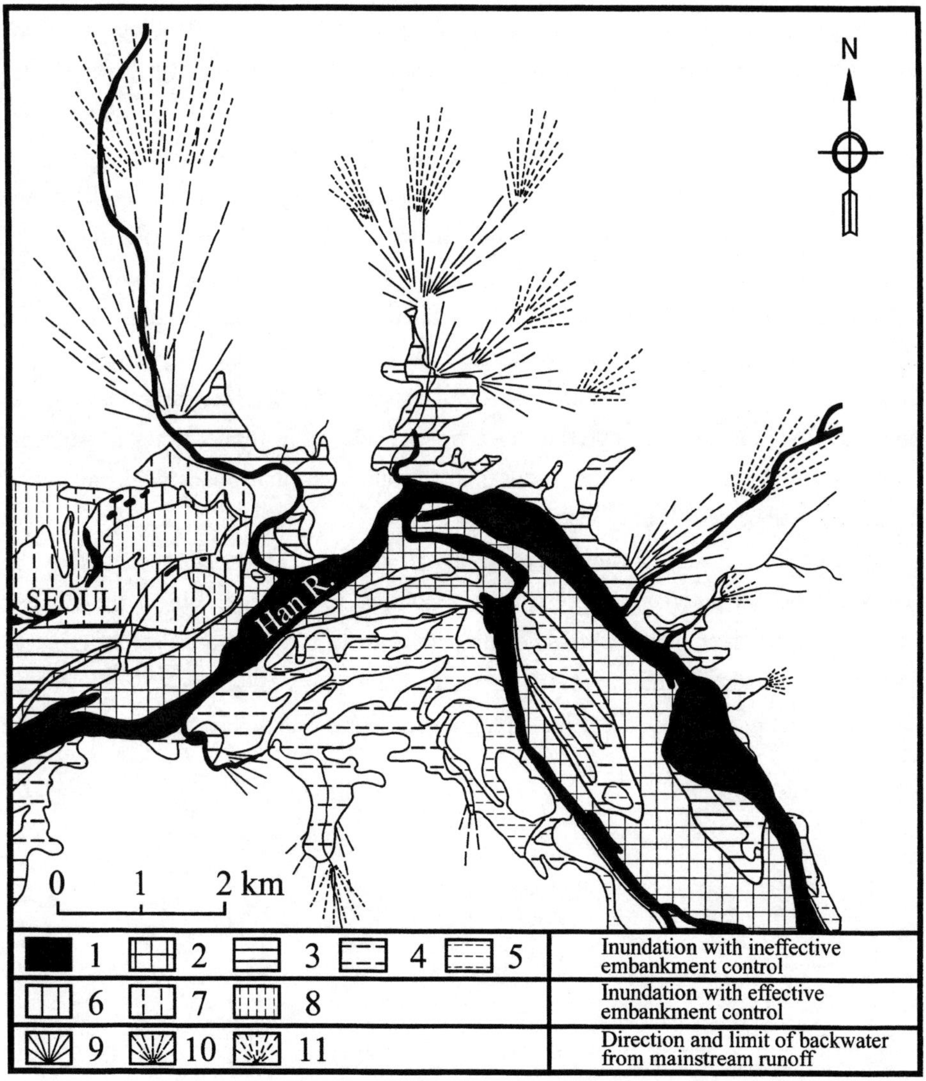

Figure 4.12 Stages of inundation in the lower reaches of the Han River (Kim, 1974; modified)
(1) River (2) First stage: River nears capacity, (3) Second stage: River overflows banks, (4) Third stage: Inundated area based on August 1972 flood (50-year flood), 5) Fourth stage: Inundated area during super-scale flood, (6) First stage of inundation with effective embankment control, (7) Second stage of inundation with effective embankment control, (8) Third stage of inundation with effective embankment control, (9) Area affected by backwater during flood, (10) Area affected by backwater during large-scale flood, (11) Area influenced by backwater during super-scale flood.

with the map of inundation stages shows close relationships between geomorphology and inundation. Kim divided the stages into two categories: inundation with ineffective embankment control and inundation with effective embankment control.

Furthermore, he showed the direction and limits of the backwater flooding from the mainstream runoff. He divided the regions affected by backwater flooding into three areas depending on the level of flooding. The map shows clearly the overflowing flood type. The central part of Seoul is located at the foot of the lower part of the mountain and piedmont slope because that area is free from flooding and deposition.

Although the amounts of recent deposition on the alluvial plains of the Han River are large, with one flood bringing from 1 to 3 meters, in ancient times depositions were probably more moderate.

Along the Han River near Seoul there is a village called Pungnabri. About 2,000 years ago, when the region was known as Baekje, a castle was built along the riverbanks. The lower edge of the castle is now buried by about 5 meters of sand. This part of the castle is only 1 to 2 meters above the present river water level, but in former times floodwaters probably did not reach the castle. From this it would seem that floods were small, and hence depositions minor, until recent times.

The recent increases in the amount of deposition of sand and gravel transported by the rivers can be attributed to deforestation. According to a gauge pole in the Bukhan Mountain, surface erosion from the mountain amounts to 15 centimeters per year.

Unfortunately, the natural levee that contained the remains of the castle was destroyed to develop the grounds for the 1992 Olympics in Seoul. The author believes it was the highest natural levee in the world.

Photo 4.1 The Bukhan Mountains near Seoul are comprised of granite. (1970)

4.10. Map Utilization for Deciding Embankments along the Mekong River

The Mekong River rises in the Tibetan Plateau, and flows through a gorge across the Unnan District of China. After that, the river runs along the border dividing Myanmar (Burma), Laos and Thailand. Finally, the river runs through Cambodia and Vietnam to enter the South China Sea. Hence, the river is international, with its basin including six countries (*Figure 4.13*).

A master plan for development of the river was prepared by the Mekong Committee of the Economic Commission for Asia and the Far East. Japan has participated in that project since 1960, and produced the development plan for the Mekong tributaries. The United States produced the plan for the main river.

Following the construction of a proposed dam, the state of flooding in the plain below the dam will change markedly, creating an enormous problem for the farmers living there. The Mekong Committee has used the geomorphological survey map to estimate the amount of flooding.

Under the Pa Mong Dam project, the author was asked to determine the present state of flooding in the Vientiane Plain, for which a geomorphological survey map of the

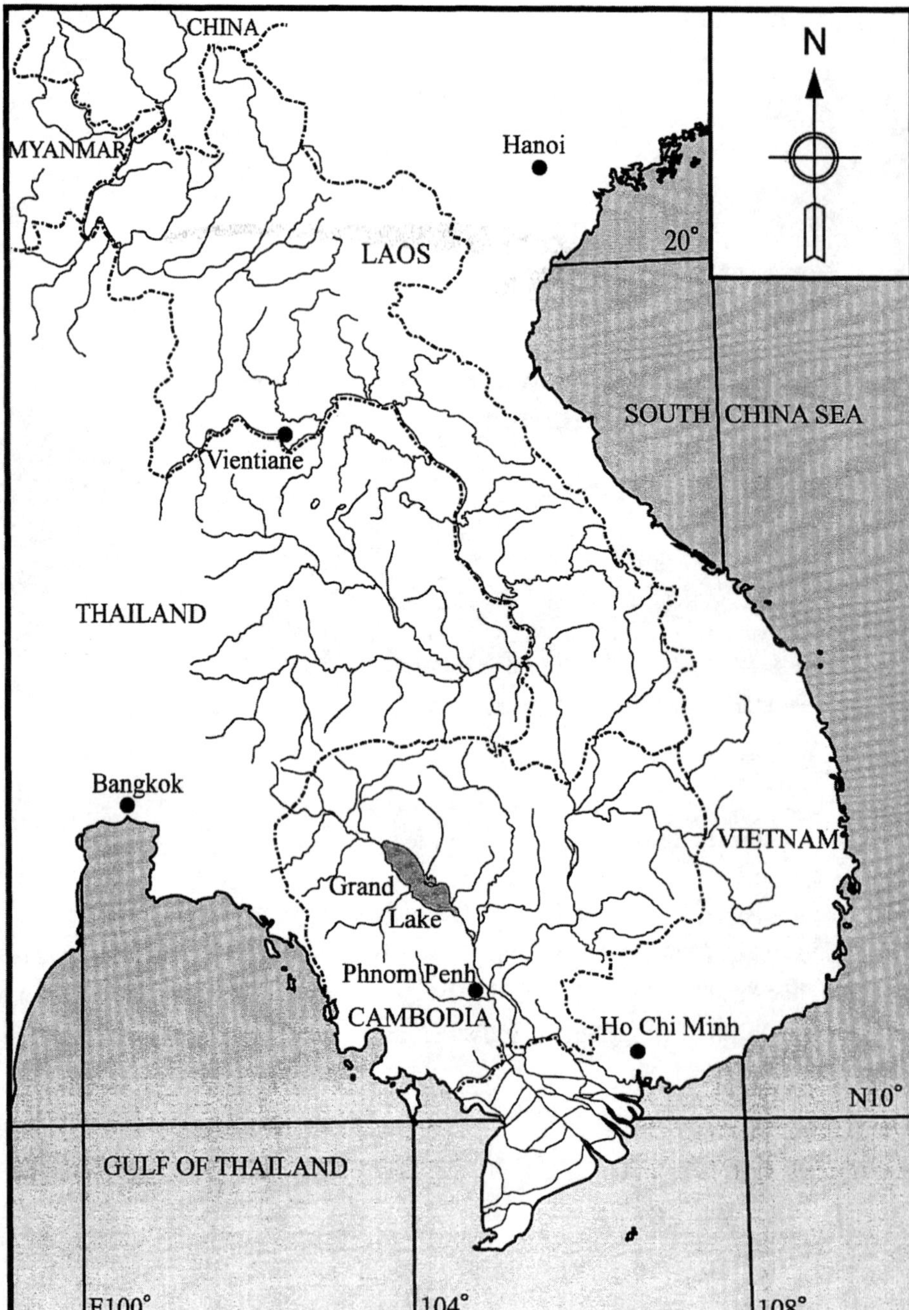

Figure 4.13 Mekong River System

Mekong River basin was prepared (*Figure 4.14*). It was observed that the alluvial fans in this region are small owing to the flatness of the topography and chemical weathering of the rocks into laterite. On the other hand, there are large natural levees along the river, and back swamps generally occupy the areas between two natural levees or between a natural levee and a terrace, and marshes are located in the terrace.

The geomorphological survey map of the Mekong River basin helped the author to estimate that 684.7 square kilometers would be the maximum area underwater during exceptional floods. The total area underwater during the 1966 flood was 499.5 square kilometers, making it the basin's largest flood of the Holocene Age so far. The over-flowing type of flooding was seen in the upper part of the Vientiane Plain, while the lower part was dominated by the concentration type (*Figure 4.15*). Hence, protective dikes are more useful in the upper part of the plain than in its lower part.

The map also helped in the construction of a railway line between Nongkhai and Vientiane and in avoiding flood-prone areas for the selection of station sites (Oya, 1967, 1980).

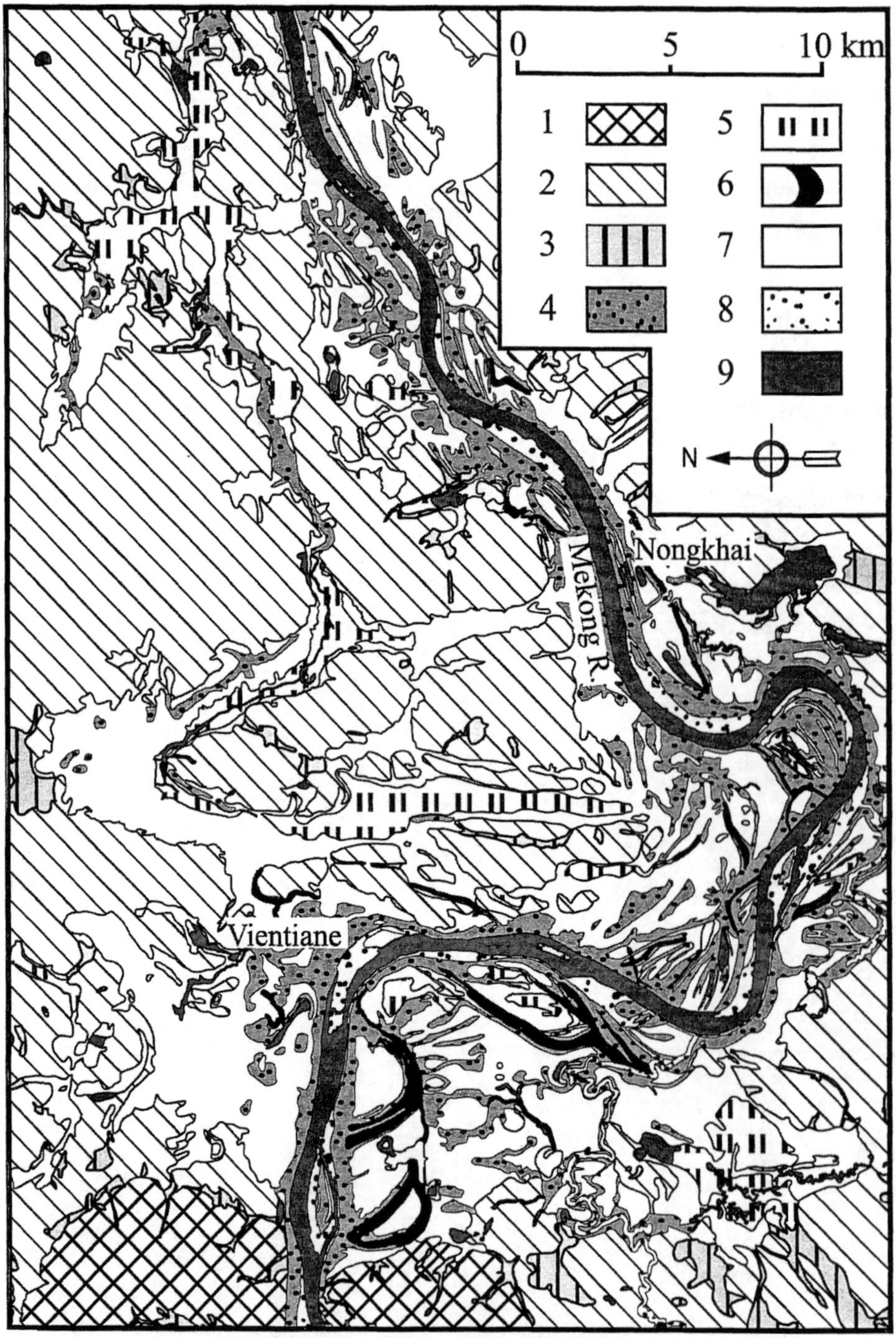

Figure 4.14 Geomorphological land classification map of the Vientiane Plain, Mekong River (Oya, 1967; modified) (1) Mountain, (2) Terrace, (3) Valley plain, (4) Natural levee, (5) Marsh, (6) Former river course, (7) Back swamp, (8) Dry riverbed, (9) Area submarged during dry season.

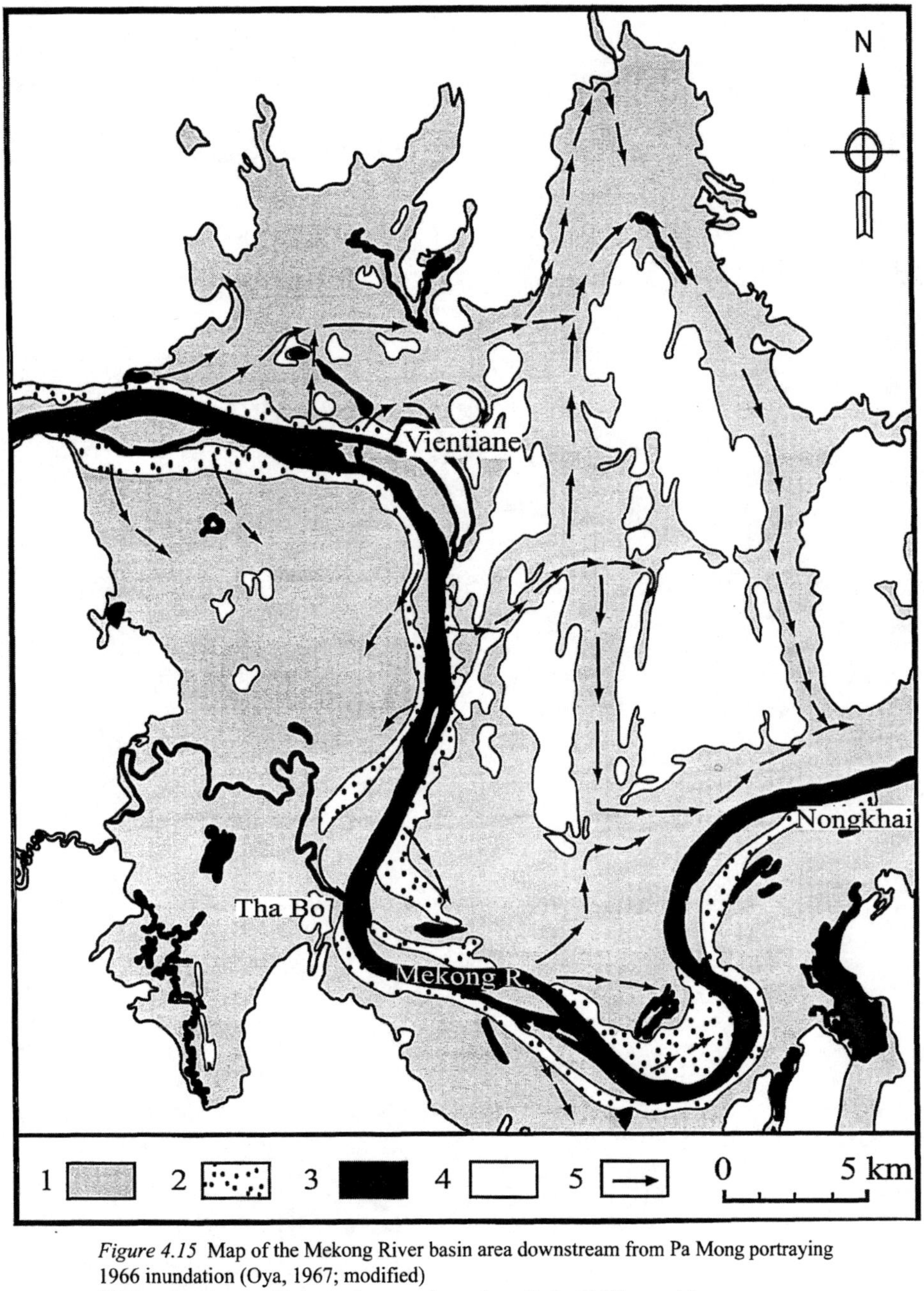

Figure 4.15 Map of the Mekong River basin area downstream from Pa Mong portraying 1966 inundation (Oya, 1967; modified)
(1) Inundated area, (2) Area where sand was deposited, (3) River or lake,
(4) Temporary dike, (5) Flood current direction.

CHAPTER 5

FLOODING IN SEMI-FRIGID ZONES

Hokkaido, Japan's northernmost main island, is situated between 41° N to 46° N and its climate is marked as Df in the Köppen classification (*Figure 4.2*). Its rivers have different characteristics from those in Japan's other main islands of Honshu, Shikoku and Kyushu, which are classified as Cf.

The author wishes to discuss the features of river flooding and topography of the fluvial plains in semi-frigid zones, using rivers in Hokkaido and Poland—which is also in the Df zone—as examples.

5.1. The Ishikari River

The Ishikari River, the largest river in Hokkaido, rises in Mount Ishikari (1,980 meters) in the central part of Hokkaido, rounds the Taisetsu Volcanic Mountains, and then flows down the Sounkyo Canyon to the Kamikawa Intermontane Depression. It then runs through the Kamuikotan Canyon to the Ishikari Plain. The Uryu River joins it near the city of Fukagawa, the Sorachi River near the city of Takikawa, the Yubari and Chitose rivers near the city of Ebetsu, and the Toyohira River near Sapporo. It then pours into the Japan Sea.

The length of the river is 286 kilometers, its drainage area is 14,330 square kilometers and its maximum flood discharge is 12,697 cubic meters per second at the Ishikari Bridge. From a geomorphological viewpoint, the river is divided into the upper mountainous region and lower Ishikari Plain by the Kamuikotan Canyon (*Figure 5.1*).

There are two causes of flooding in the Hokkaido district: the melting of snow in spring and the typhoons that occur in summer.

The Ishikari River has greater levels of flooding in summer than spring. Its slope is gentler than the slopes of the Tone and Yamato rivers in Honshu (*Figure 5.2*), and the velocity of its flood discharge is slower than those rivers, resulting in less flood damage. However, the volume of soil coming from the mountainous district is large, due to the great amount of soil erosion caused by snowmelt.

The combination of geomorphic elements in the Ishikari Plain is as follows:

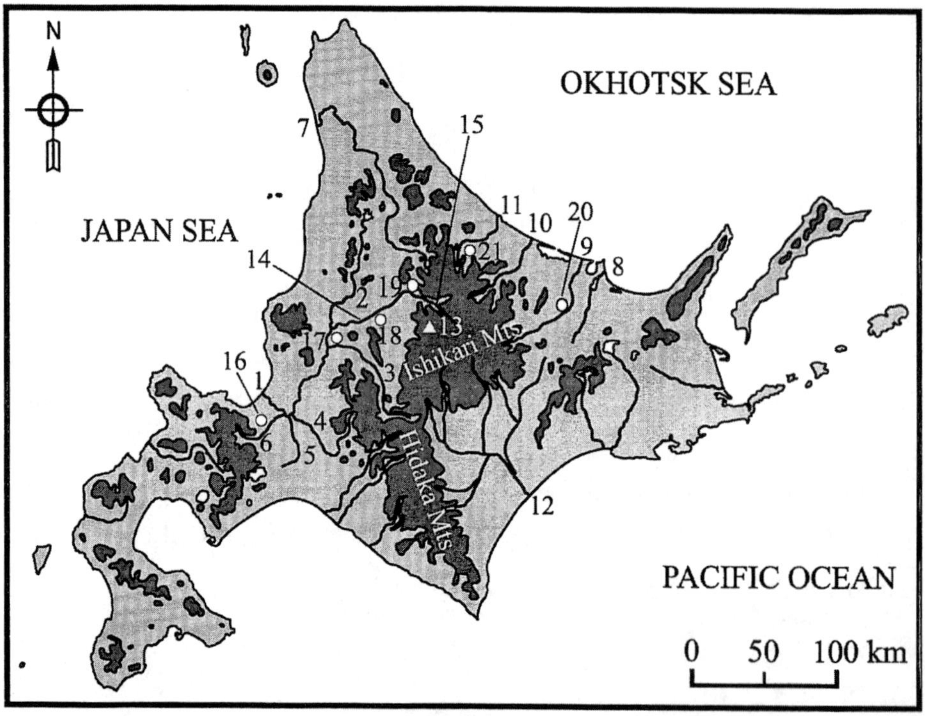

Figure 5.1 General topography of Hokkaido
(1) Ishikari River, (2) Uryu River, (3) Sorachi River, (4) Yubari River, (5) Chitose River, (6) Toyohira River, (7) Teshio River, (8) Abashiri River, (9) Tokoro River, (10) Yubetsu River, (11) Syokotsu River, (12) Tokachi River, (13) Taisetsu Volcanic Mountains, (14) Kamuikotan Canyon, (15) Sounkyo Canyon, (16) Sapporo, (17) Takikawa, (18) Asahikawa, (19) Kamikawa, (20) Kitami, (21) Takinoue.

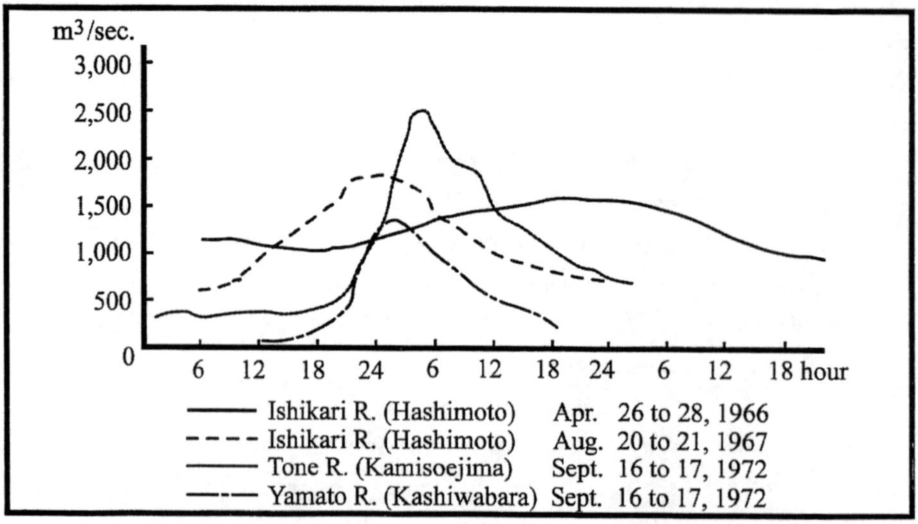

Figure 5.2 Comparison of the hydrograph record of Ishikari River flooding with those of the Tone and Yamato rivers (Oya, 1981; modified)

Fan + Natural Levee + Natural Levee with Peat +Delta

↑

Toyohira River's Fan

Large natural levees, two to five kilometers wide, lie along the main Ishikari River, and there are back-marshes between the natural levees and mountains. Peat-bog layers are well developed in these back-marshes because of the island's cold climate.

The plain's inundation shows the following unique features: The peat in the area, which is deposited on clay, forms high, intermediate and low moors. The high moors in the central part of the back-marsh area are surrounded by rings of intermediate and low moors (*Figure 5.3*). The maximum depth of the peat-bog layers is 6 meters.

The author considers that before the peat developed in the back-marsh area, the floodwater of the Ishikari River crossed the natural levee into the back-marsh area and inundated it for a long period. The surface of the back-marsh gradually rose due to the deposition of peat, and finally reached the level of the natural levee.

Because of this high elevation, the high moor areas are now safe from flooding. Furthermore, floodwater from branch courses becomes concentrated and flows into the main course of the Ishikari River. Thus, unlike the rivers of Honshu in which floodwater disperses from natural levees into back-marsh areas, Ishikari floodwater flows down at a concentrated two to five kilometers width on the natural levees.

The Ishikari Plain was devastated by floods during the periods of Aug. 15 to 18, 1932; April 9 to 12, 1955; April 15 to 20, 1956; and Aug. 3 to 5, 1981. During all those floods the high moor areas were safe from invasion, particularly the high moor area of the Shinotsu Plain (*Figure 5.3*). In the 1981 flood, although the peat bogs had been converted to paddy fields or to upland crops fields, the above mentioned features of flooding in peaty areas were still seen.

5.2. Rivers and fluvial Plains along the Okhotsk Sea

Rivers and fluvial plains along the Okhotsk Sea lie in the coldest region of Japan. The Abashiri, Tokoro, Yubetsu and Shokotsu rivers run from south to north.

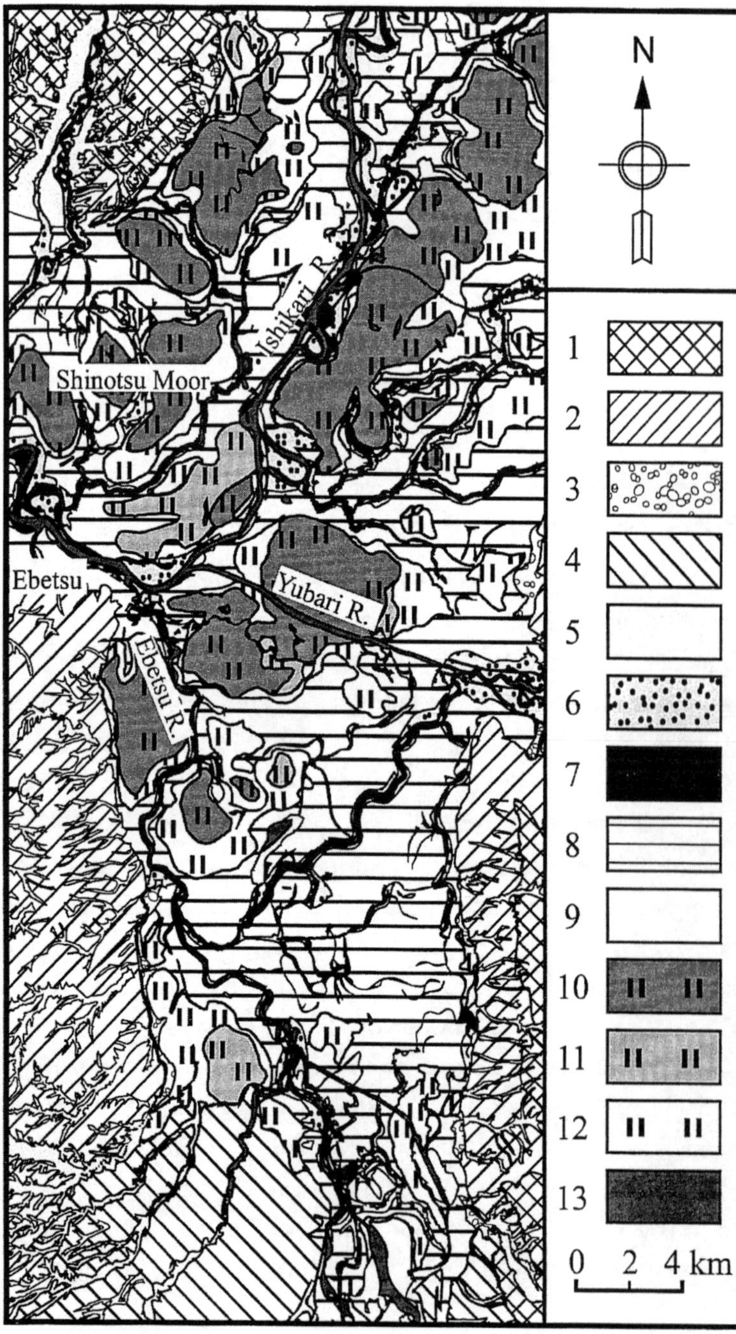

Figure 5.3 Topographical survey map of the Ishikari River basin (Oya, 1959; modified)
(1) Mountain, (2) Terrace, (3) Fan, (4) Gentle slopes at the foot of the volcano, (5) Valley plain, (6) Natural levee, (7) Former river course, (8) Delta or back-marsh, (9) Reclaimed land, (10) Higher moor, (11) Intermediate moor, (12) Lower moor, (13) River.

The combinations of geomorphic elements of the fluvial and coastal plains of the area are as follows:

Abashiri River: Natural Levee + Delta
Tokoro River: Alluvial Terrace + Natural Levee + Sand Dune
Yubetsu River: Fan + Delta + Sand Dune
ShokotsuRiver: Fan + Sand Dune

Based on these combinations of geomorphic elements, the fluvial and coastal plains along the Okhotsk Sea can be divided into two groups: Group A includes the Abashiri and Tokoro rivers, and Group B the Yubetsu and Shokotsu rivers.

5.2.1. FEATURES OF GROUP A

Between the terraces and mountains are gentle slopes that consist of clay and angular gravel. It is believed that the geomorphology was formed by solidification under a colder climate, perhaps the Würm Ice Age. At that time, the area was Peri-glacial. There is a danger of landslides when construction projects, such as the building of roads, are undertaken.

The gradient of the plain is gentle and there is no fan. Here, natural levees are only developed along the main river courses. This contrasts with Honshu, where natural levees exist not only along the main rivers but also along their distributaries. For example, in the Nobi Plain, there are many natural levees along the distributaries and abandoned river channels of the Kiso River (*Figure 3.1*), indicating that changes in the river course were frequent. The natural levees are large, and are high relative to the back-marshes. In comparison, the relative heights of the natural levees along the rivers of Group A are low. This is because peat is developed in the back-marshes and flooding from melting snow is moderate.

The shifting of the Tokoro River course has been considerable, especially from west to east near the town of Tanno, which lies in the lowest reach of the Kitami Intermontane Depression. The author attributes the river's shift to freezing and liquefying of the soil.

A narrow valley lies between the Kitami Intermontane Depression and Tokoro Plain, and in the spring, ice jamming sometimes causes flooding at the lower edge of the depression.

5.2.2. FEATURES OF GROUP B

The gradient of the plain is steep. A fan is developed at the outlet of the wide canyon but there are no natural levees. A small delta is formed on the inland side of the coastal sand dune, and a thin peat-bog layer is present.

Since these river basins are located in high latitude, between 43.5°N and 44.5°N, the area belonged to the Peri-glacial region during the glacial ages. A large amount of sand and debris from weathering and frost action in the mountainous region was deposited in the valley and at the foot of the mountains. Therefore, the mountains are shaped as in an old stage, and the piedmont gentle slope is well developed.

Land collapse is minor in the upper reaches of these rivers, but there is much gravel in the lower reaches. At present, the rivers are eroding this sand and debris that was deposited in the glacial age, and are transporting it to the lower reaches where many angular types of gravel can be seen.

The annual precipitation in the Okhotsk coastal region is low; the average annual rainfall at the city of Kitami is about 700 millimeters. The rainfall is highest from June to October.

Floods here have two causes: a) melting of the snow in spring and b) torrential rainfall caused by typhoons in summer. A flood caused by a) is moderate but lasts a long time; discharge is high and a lot of silt is transported because the surface of the ground has been eroded by the melting snow. In spring, ice jams sometimes cause flooding. A flood caused by b) is rapid but its duration is short. Sand and gravel are transported.

5.2.3. COMMON FEATURES OF A AND B

Hokkaido receives a lot of snow in winter, with the town of Takinoue getting as much as 250 to 300 centimeters annually. However, the amount varies regionally. In areas with less snow, the surface soil freezes readily. When spring comes, the frigid soil layer near the surface begins to melt but the deeper parts remain frozen. Water cannot penetrate this frozen layer. Therefore the soil layer near the surface is saturated by the melting snow and ground water, and flows down the mountain slopes easily. Generally speaking, mountain slopes in the Hokkaido region are eroded horizontally by this melting snow. Fans here consist of gravel and mud that was produced in this manner.

5.3. The Vistula River in Poland

The Vistula River rises in the Tatra Mountains, which lie between Poland and the Czech Republic, and runs east to northeast through Krakow. It turns north at Sandomùrz, and then to the west at Warsaw. After Torun, the river runs northward and pours into the Bay of Gdansk in the Baltic Sea (*Figure 5.4*).

The author surveyed the river and plain using landsat images and field surveys, in which he was assisted by the Polish Academy of Sciences.

Much of the Vistula River from Warsaw to the lower reaches runs from east to west, parallel with the coast of the Baltic Sea. This direction shows that the rivers ran along the south edge of the morainic plateaus that were formed during the Leszone, Poznan and Pomeranian phases in the last glacial age (*appended Table 3*). Namely, the direction of the rivers was determined by the advance of the glacier coming from the Baltic Sea.

According to Jozef Mojski (1980), the sand and gravel layer situated on the marine deposits of the Eemian Interglacial Age is divided into five zones. Deposits from the last glacial age have eroded away, and the age of the lowest layer is on the boundary between late Pleistocene and oldest Holocene, or about 10,000 years B.P. The Holocene beds are about 25 meters thick. The author could spot a minor glacial and interglacial age within the layers.

The Vistula Plain is divided into the following geomorphic elements: fan, natural levee, lagoonal lowland, delta, sand spit and sand dune. There is no alluvial fan along the main river, but one is present at the foot of the morainic plateau on the western edge of the plain (*Figure 5.5*).

Natural levees are well developed along a tributary that flows through the town of Nogat to Vistula Lake, and along a branch that reaches the city of Gdansk. Back-marshes are developed between the natural levees.

There are three lines of sand spits along the seacoast, which have been converted into sand dunes by the strong northwest wind. In these dunes, the author found three layers of fossil soils, indicating changes of climate.

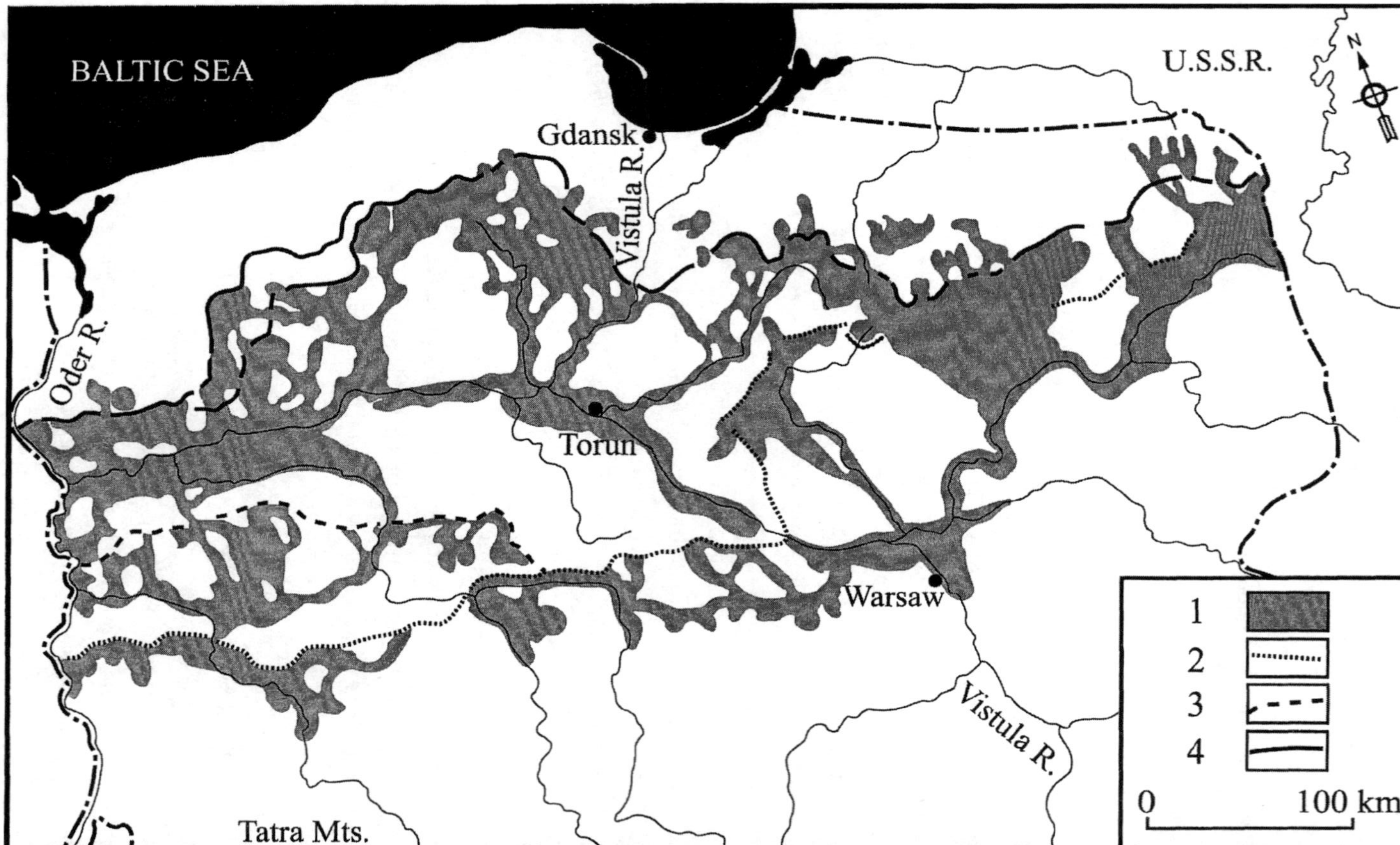

Figure 5.4 Map of the Vistulian in the Polish Lowlands (Mojski, 1980; modified)
(1) Glaciofluvial valley, (2) Southern limit of Leszno Moraine, (3) Southern limit of Poznan Moraine, (4) Southern limit of Pomeranian Moraine.

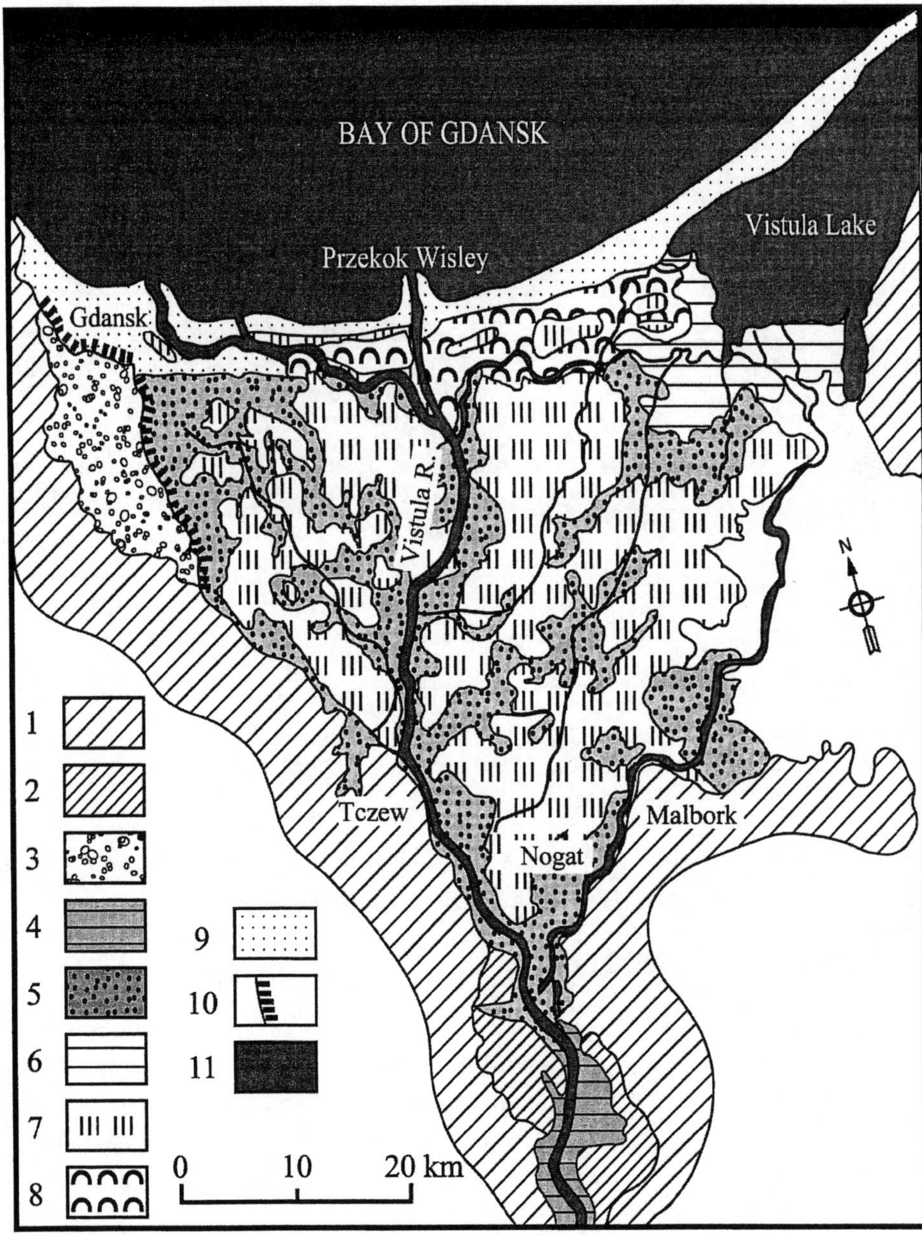

Figure 5.5 Geomorphological classification map of the Vistula River in Poland (Oya, 1987; modified) (1) Morainic terrace, (2) Terrace, (3) Fan, (4) Valley plain, (5) Natural levee and higher delta, (6) Former lagoon and delta, (7) Back-marsh, (8) Sand dune, (9) Sand spit, (10) Cliff formed by Litorina Transgression, (11) River, lake or sea.

Photo 5.1 Vistula River in Warsaw, Poland (1987)

Date	Water Level
Feb. 18, 1570	10.74 m
Feb. 17, 1584	10.56 m
March 16, 1719	10.21 m
March 1, 1871	9.78 m
Feb. 18, 1879	9.86 m
March 13, 1891	9.80 m
Aug. 2, 1980	8.00 m

Table 5.1 Record of flooding at Torun

The hydrograph at Tczew shows that almost all the flooding occurs during the spring (*Figure 5.6*). The Tatra Mountains are covered with about 2 meters of snow in winter. In spring, warm south winds melt the snow, which along with ice jams, causes floods.

About 100 years ago, the water level of the Baltic Sea rose due to blocking by ice jams, resulting in the erosion of the coastal sand dunes between Gdansk and the present river mouth. Immediately afterward, the Swibno Flood Way was constructed to protect Gdansk from flooding. The flood way, later renamed Przekok Wisley, then became the main course of the Vistula. Construction of the flood way increased the gradient of the Vistula and flood discharge increased (*Table 5.1*).

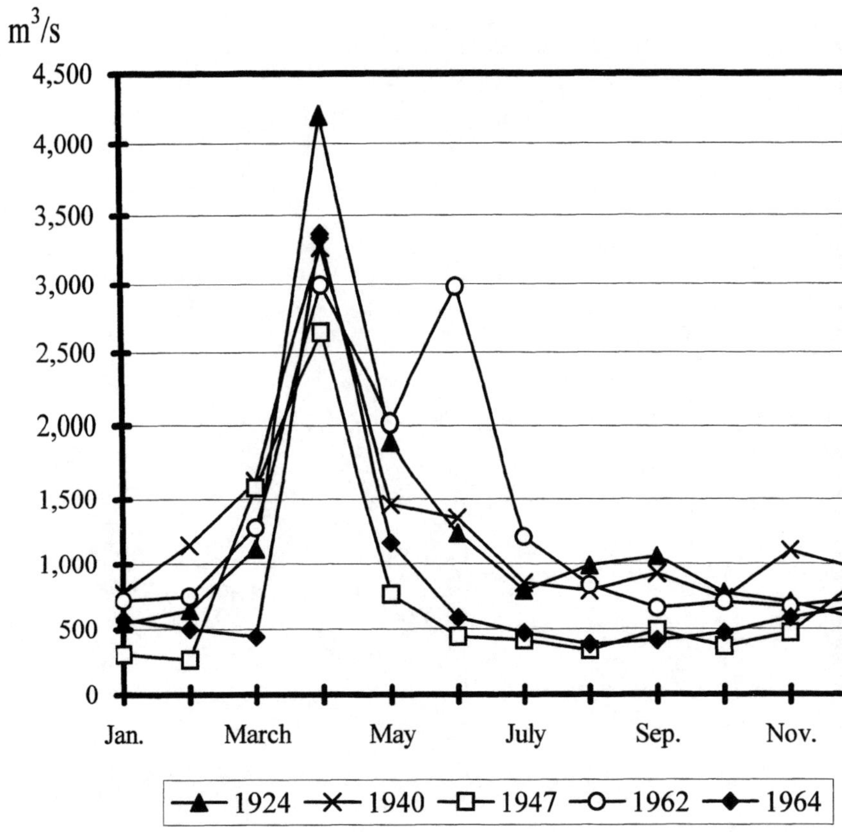

Figure 5.6 Hydrograph record of flooding at Tczew (Mojski, 1980; modified)

Between 1895 and 1965, Przekok Wisley was extended by a jetty about 2.2 kilometers in length. Blocking of the river mouth by ice jams and storms continues to cause rises in the water level.

As mentioned, flooding occurs when the development of sand spits block river mouths. To lessen flooding, flood ways cutting coastal sand dunes have been constructed along the Agano, Shinano and Mogami rivers in Japan, all situated along the Japan Sea coast. The similarities in flooding features and prevention between Japan and Poland are significant.

CHAPTER 6

DIFFERENCES IN GEOMORPHOLOGY AND FLOODING BETWEEN THE LEFT AND RIGHT BANKS OF RIVERS ALONG THE MEDIAN DISLOCATION LINE

6.1. Japan's Landforms

Situated in the Circum Pacific Orogenic Belt, the Japanese archipelago has significant crustal movement.

From a geological viewpoint, the archipelago can be divided into six parts. Southwestern and northeastern Japan are separated by the Fossa Magna, southwestern Japan is divided into the Inner Zone (on the Japan Sea) and the Outer Zone (on the Pacific Ocean) by the Median Dislocation Line, northeastern Japan is divided into the Inner Zone (on the Japan Sea) and the Outer Zone (on the Pacific Ocean) by the Morioka-Shirakawa Line, and Hokkaido is divided into the Inner Zone (Oshima Peninsula) and the Outer Zone (mainland of Hokkaido) by the Sapporo-Tomakomai Line (*Figure 6.1*).

The Fossa Magna region was a depressed area in former times but is currently covered by volcanoes, including Fuji Volcano. The crustal movement in the Inner Zone of southwestern Japan is of a block movement type, with each block limited by fault lines. There are many intermontane depressions among the horst mountains. While there are few plains in this area's Outer Zone, there are many in its Inner Zone.

We shall discuss the regional differences between the plains of the Inner and Outer zones of southwestern Japan, using the Yoshino River in Shikoku and the Toyo River in the district of Chubu as examples. Both these rivers run along the Median Dislocation Line.

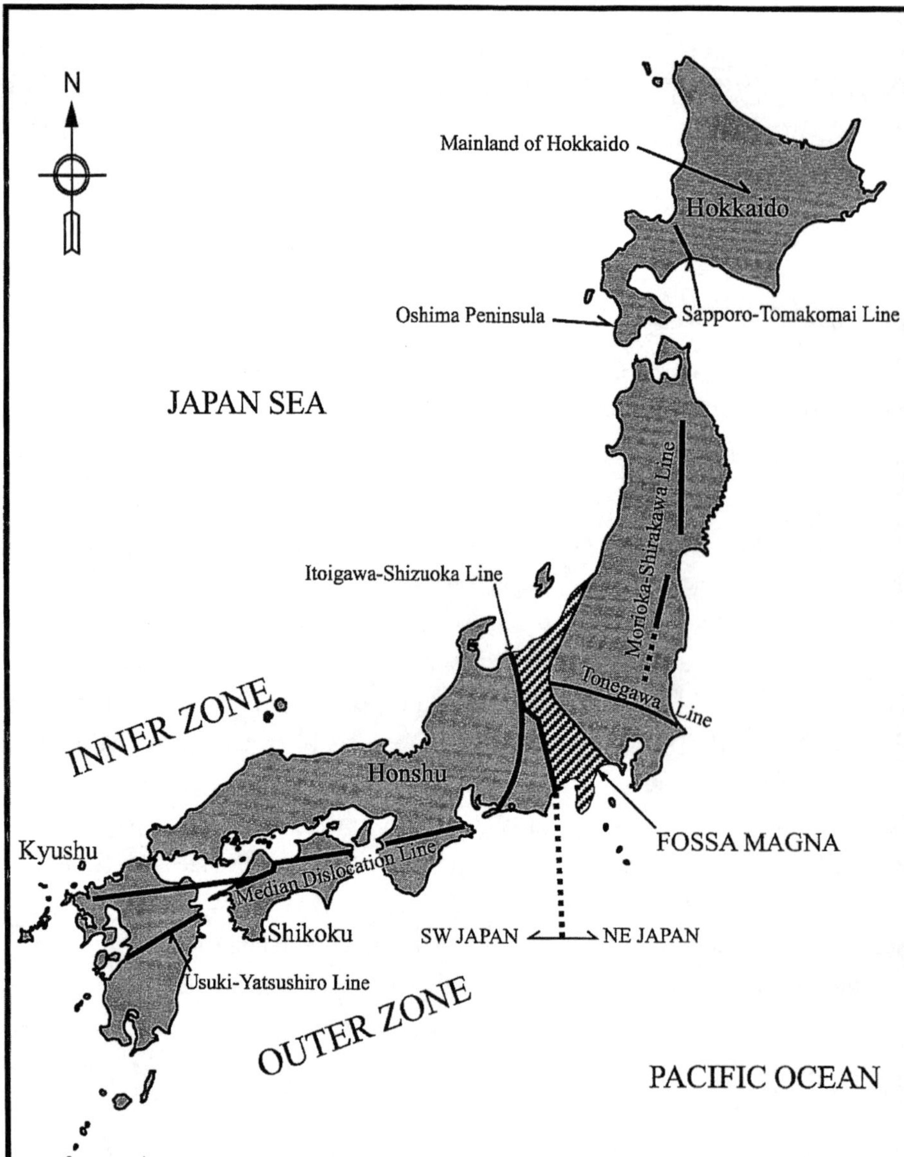

Figure 6.1 Geological structure of Japan

6.2. The Yoshino River

6.2.1. BRIEF DESCRIPTION OF THE YOSHINO RIVER

The Yoshino River rises in the western part of the Shikoku Mountains, runs east and then turns north. It crosses the Shikoku Mountain Range through ravines, called the Oboke and Koboke gorges. From a geomorphological viewpoint, the gorges formed in an antecedent valley. In these gorges, the river joins with the Iya River from the east and the Dozan River from the west. After leaving the Shikoku Mountains at the town of Ikeda, the river runs along the Median Dislocation Line, forming a narrow plain that is divided into two by the Iwazu Narrows, located 40 kilometers from the river mouth. From the Narrows to the river mouth lies Tokushima Plain, in which the river joins the Sadamitsu, Anabuki and Akui rivers from the south (Outer Zone) and the Soe-dani, Higai-dani and other rivers from the north (Inner Zone).

The Yoshino River is 194 kilometers long and its drainage basin encompasses 3,750 square kilometers.

More than 3,000 millimeters of rain falls annually in the upper reaches of the river, and the area experiences frequent typhoons. The estimated flood discharge at the Iwazu Narrows is 24,000 cubic meters per second, one of the highest levels in Japan.

6.2.2. HISTORICAL RECORDS OF THE YOSHINO RIVER

During the Edo Period, the Yoshino River was controlled by feudal lord Hachisuka. At that time, the Bekku River flowed in the Tokushima Plain. For navigational purposes, Hachisuka constructed a channel in 1672 to connect the Yoshino with the Bekku River at the Daiju Weir. Because the gradient of the new channel was higher than that of the Yoshino, it became the main course and the Yoshino became a branch.

At that time, flooding was frequent in the Tokushima Plain, and it was especially massive in 1701. Hachisuka encouraged the cultivation of indigo, as the plants would be ready for harvesting during nonflood periods. Forests, mainly bamboo, were used to protect areas from flooding in those days.

Following the Meiji Restoration, when the government began taking a good look at such matters as flood control, the Yoshino River flooded numerous times. The government began its first improvement project in 1907, constructing an embankment that took

20 years to complete. With its construction, the flood protection forest disappeared between the river mouth and the Iwazu Narrows, at the lower reaches of the Yoshino River. Forests still exist in the middle reaches, between the Iwazu Narrows and Ikeda (*Figure 6.3*). Of course, forests cannot completely protect an area from flooding, but they do help in retarding floodwater.

6.2.3. GEOMORPHIC FEATURES OF THE YOSHINO RIVER PLAIN

There are great topographical differences between the regions on the left and right banks of the Yoshino River Plain, particularly in their terraces and fans (*Figure 6.2*). While there are many Pleistocene terraces and fluvial fans on the left bank (Inner Zone), there are few on the right (Outer Zone). These terraces were originally fluvial fans that were turned into terraces by river erosion and ground upheaval (Ichinose, 1966, 1977).

The drainage basins of rivers coming from the Asan Mountain Range (Inner Zone) are smaller than those of rivers coming from the Shikoku Mountain Range (Outer Zone). But, strange to say, the former has many fans and terraces while the latter has few. They are especially well developed between the Nomura-dani and Miyagouchi-dani valleys, in the central part of the Yoshino River Plain.

Flooding on the left bank is much more dramatic than that of the right, the velocity of the current is faster, deposition of sand and gravel is larger, and the tributaries change course. Such flooding occurred in 1961, when Muroto Typhoon No. 2 destroyed the Miyagouchi-dani and Soe-dani embankments and the fan was inundated.

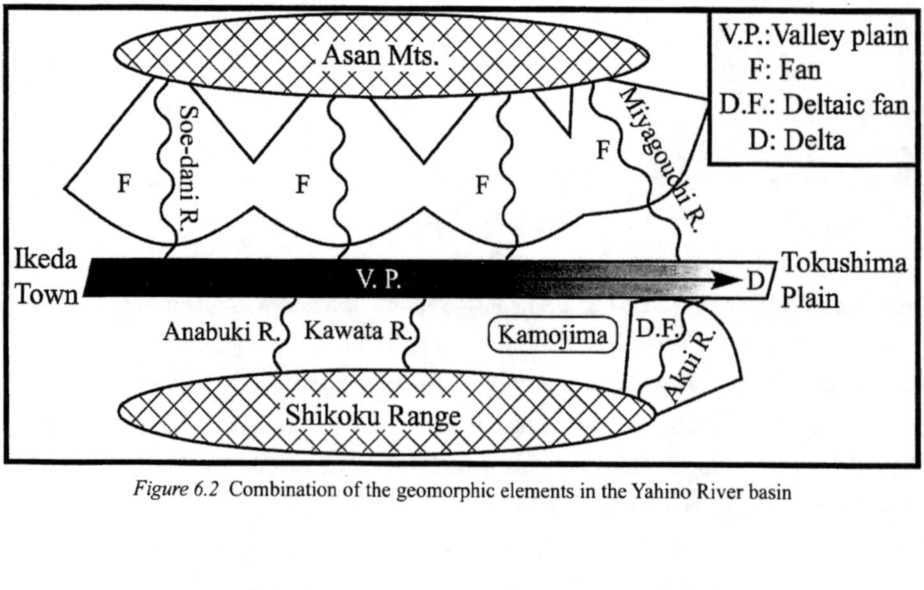

Figure 6.2 Combination of the geomorphic elements in the Yahino River basin

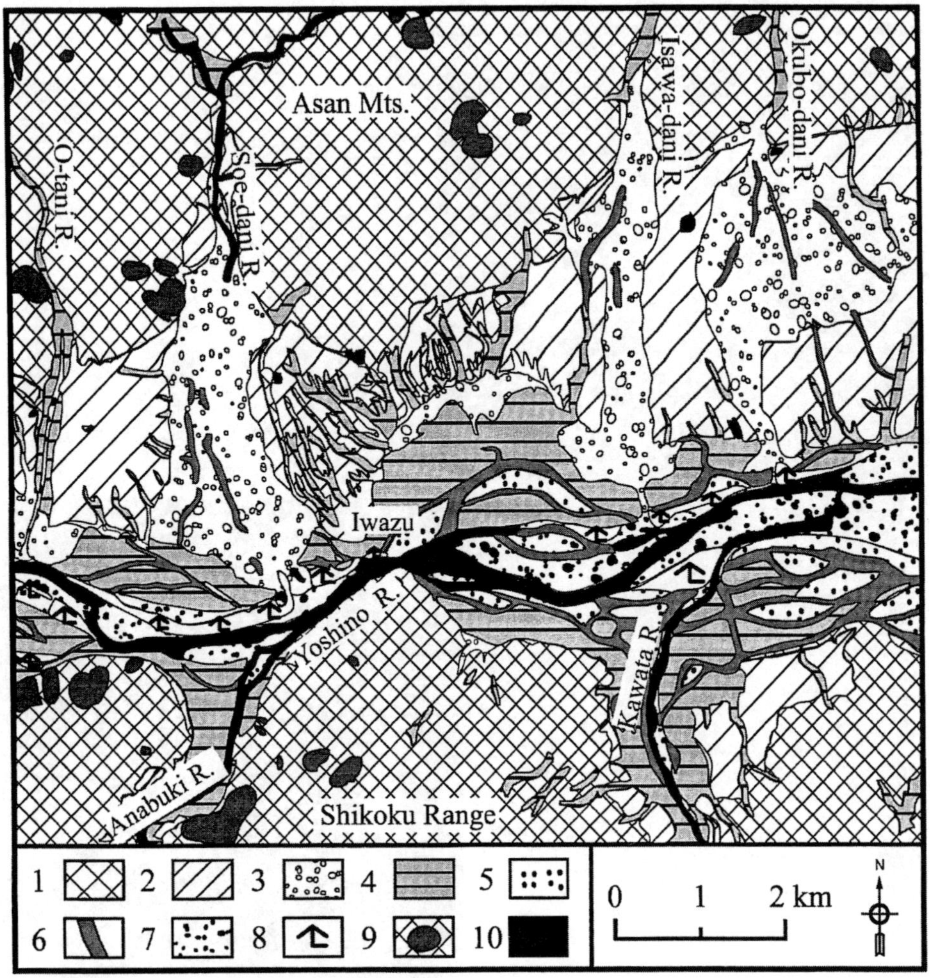

Figure 6.3 Geomorphological survey map of the middle reaches of the Yoshino River basin, Shikoku (Oya et al., 1995; modified)
(1) Mountain, (2) Terrace, (3) Fan, (4) Valley plain, (5) Natural levee, (6) Former river course, (7) Slightly hilly area formed by dry riverbed, (8) Forest for mitigation of flooding, (9) Land collapse, (10) River.

Photo 6.1 Upper reaches of the Yoshino River in the Outer Zone (1982)

Photo 6.2 A woman shows a watermark fron the Shimanto River in the Outer Zone. (1984

Photo 6.3 Farmland in the Inner Zone of Shikoku (1984)

The slopes of the Asan Mountain Range are steep, and the sandstone and shale that form the mountain are fragile. In particular, the middle of the range is high and erosion is significant.

The longitudinal profiles of the rivers are steep, with many knick points. There are large amounts of sand and gravel in their upper reaches but, because of the small flood discharge, these rivers cannot carry the material to the main course of the Yoshino River, and it is deposited at the foot of the mountains as alluvial fans. Currently, the riverbeds in the alluvial fans are elevated along artificially fixed river courses.

In contrast, few terraces and fans are present along the Kawata and Anabuki rivers rising from the Shikoku Mountains in the Outer Zone.

Earth movement in the Shikoku Mountains is of a broad warping type, and is less violent than in the Asan Mountains in the Inner Zone. Erosion is insignificant. Because the Shikoku Mountain Range is composed of crystalline schist, the formation of fine-grained materials derived from efflorescence, such as weathered clay, is important. Therefore, there is much clay but minor amounts of sand and gravel in the Outer Zone.

6.2.4. UTILIZATION OF GEOMORPHOLOGICAL LAND CLASSIFICATION MAPS

Forecasting River Flooding

A geomorphological land classification map helps us predict the features of flooding based on the combination of geomorphic elements, the flood types of which have already been shown in *Table 1.2.*

As previously stated, there are many fans in the Inner Zone but very few in the Outer Zone.

In the Inner Zone, riverbeds are higher than those in the adjacent fluvial plain, and when massive flooding occurs, the water will overflow into the adjacent lowlands.

In the Outer Zone, fans and natural levees are rare. There, a cross section shows that the land surface lowers toward the river. Therefore, during a flood, the water does not disperse but is concentrated into the lower reaches of the rivers.

A special type of flooding occurs in the eastern part of the district of Kamojima, in the Outer Zone. The slightly hilly area along the abandoned rivers trend west to east, parallel with the Yoshino River and its dike, making drainage difficult. Furthermore, the deltaic fan formed by the Akui River has developed from south to north in the eastern part of the district, also making drainage difficult. Inundation is long and deep (*Figure 6.4*).

Tsunami Damage Along the Seacoast and Delta Region

There is much recorded evidence of the many earthquakes that have occurred along Japan's Pacific Coast, which have caused frequent damaging tsunami.

The Ansei Earthquake of 1854 caused a tsunami that invaded into the delta region via the former Yoshino River. Several stone monuments have been erected along the former Yoshino in memory of those who died as a result of tsunami. The present river course, former river course and canal are convenient routes for tsunami invasion.

Geomorphological land classification maps have proved useful in a number of areas. Hitoshi Murakami of Tokushima University has been using the maps to work on mathematical estimations of tsunami. As well, the Geographical Survey Institute of the Ministry of Construction has used the maps to estimate the amount of flood damage that would occur if a rise in the Earth's temperature raises the sea by a meter.

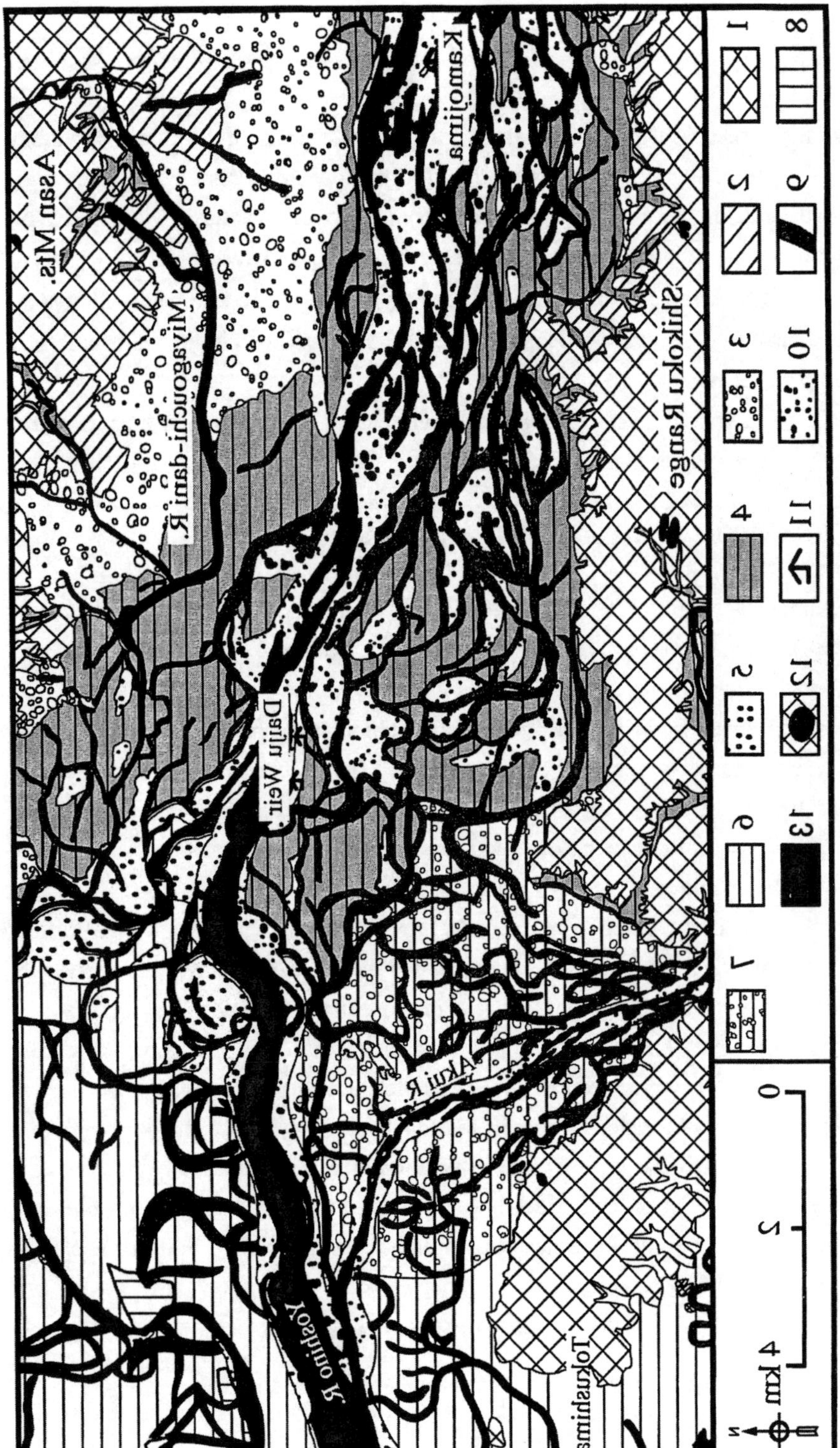

Figure 6.4 Geomorphological survey map of the lower reaches of the Yoshino River basin, Shikoku (Oya et al., 1995; modified)
(1) Mountain, (2) Terrace, (3) Fan, (4) Valley plain, (5) Natural levee, (6) Delta, (7) Artificially filled field, (8) Former river course, (9) Slightly hilly area formed by dry riverbed, (10) Forest for mitigation of flooding, (11) Land collapse, (12) River.

6.3. The Toyo River

The Toyo River rises in Danto Mountain (1,152 meters), part of the Mikawa Highland in the central part of Japan, and flows southeast to the town of Nagashino. At the town, the river changes direction to flow southwest to the city of Toyohashi via the city of Toyokawa along the Median Dislocation Line. It then enters Mikawa Bay.

Its length is 83 kilometers and its drainage area is 721 square kilometers. Its maximum flood discharge, recorded at Ishida Research Point near the city of Shinshiro, was 4,770 cubic meters per second on Aug. 5, 1969.

From a geomorphological viewpoint, the river can be divided into an upper mountainous region and a lower plain near Shinshiro.

6.3.1. GEOMORPHIC FEATURES OF THE TOYO RIVER

The Median Dislocation Line is located along the Toyo River, creating distinct regional differences in geomorphology between its right bank, which is in the Inner Zone, and left bank, in the Outer Zone (*Figure 6.5*).

Crustal movement in the Inner Zone is that of a block movement type, but the Outer Zone is characterized by warping. The rocks in the Inner Zone consist of granite and metamorphic types. Tertiary Age formations are present. On the other hand, crystalline schist is prominent in the Outer Zone, and Paleozoic Age formations are found there. Rivers in the Inner Zone are steeper but shorter than those in the Outer Zone. There are many knick points in the Inner Zone rivers, but few in those of the Outer Zone. Profiles of the Outer Zone rivers show a graded course, and the volume of gravel transported by tributaries in the Inner Zone is larger than that of those in the Outer Zone.

Terraces developed in the Inner Zone are wider than those in the Outer Zone. The terraces are divided into three groups: upper, middle and lower.

An upper terrace is developed only in the Inner Zone. Initially formed by sand and gravel brought in by tributaries, it was then dissected by ground upheaval.

Middle terraces exist on both sides but are wider on the right bank. Shinshiro and Toyokawa are located on these terraces. According to carbon dating, the marine clay in these terraces is 30,000 years old.

Lower terraces are less developed than upper and middle terraces.

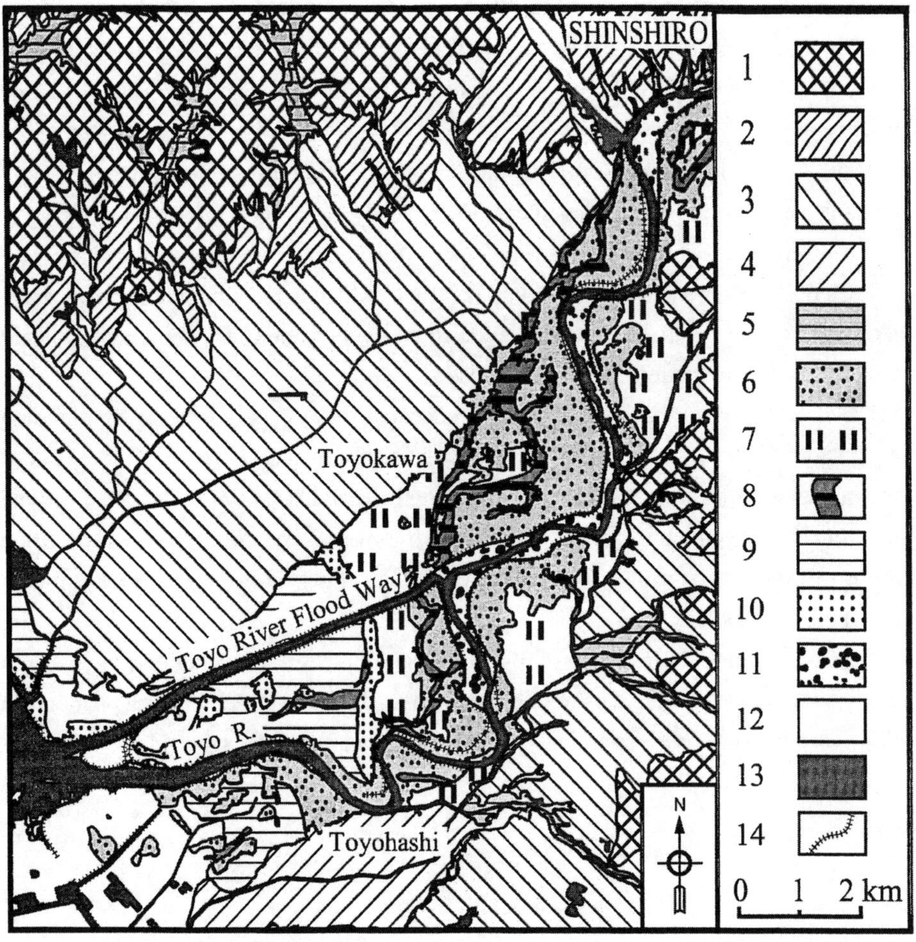

Figure 6.5 Geomorphological survey map of the Toyo River basin (Oya and Omori, 1978; modified) (1) Mountain, (2) Upper terrace, (3) Middle terrace, (4) Lower terrace, (5) Valley plain, (6) Natural levee, (7) Back-marsh, (8) Former river course, (9) Delta, (10) Sand spit, (11) Dry riverbed, (12) Reclaimed land or artificially filled field, (13) River or sea, (14) Embankment.

The terraces are never submerged during periods of flooding, but the valley floor between terraces can become submerged. The flood current in the area is fast and the water drains away well.

There is no fan in the plain despite the river's transportation of gravel. The author believes this is because the plain is too narrow for a fan.

Natural levees are well developed in the plain, with the natural levees on the right bank larger than those on the left. The relative height of the former is about 2 meters and the latter about 1 meter. The levees, which were formed during the Jomon Transgression, when the sea level was from 2 to 6 meters higher than now, reveal the courses of many former rivers.

While the natural levees can become submerged by extreme flooding, the water drains off well. However, back-marshes remain deeply submerged for long periods.

If we were to draw a line connecting Yoshida Bridge in Toyohashi to Toyokawa, we would find the lower section to be delta. This area, which is reclaimed land constructed after 1650, was devastated by a tidal wave caused by a typhoon in 1953, and it is likely that it will be further battered by high waves in the future.

The natural levees, back-marsh region and delta have experienced much flooding since medieval times, resulting in many flood-control projects. A record even exists of the construction of an open levee in 1604. Continuous dikes were constructed around 1885, but five open embankments were left for use as retarding reservoirs during periods of flooding. After the Toyo Flood Way was completed in 1965, the delta region was safe from flooding.

Thus the differences in topography and flooding between the Outer and Inner zones are apparent not only in the Yoshino River but also the Toyo River. Similar phenomena will be seen in every river running along the Median Dislocation Line.

CHAPTER 7

ESTIMATION OF LAND COLLAPSE IN JAPAN'S MOUNTAINOUS AND VOLCANIC REGIONS

7.1. Land Collapse in Mountainous Regions

7.1.1. FEATURES OF MOUNTAINS AND VOLCANOES IN JAPAN

Japan's mountainous regions (*Figure 7.1*) were formed by upheaval and plains by depression. *Figure 7.2* shows the vertical displacement of Japan during the Quaternary Period. Regions were thrust upward, particularly on the island of Honshu.

At present, the Japan Alps rise about 3,000 meters above sea level, but they were only about 1,000 to 1,500 meters high by the end of the Tertiary Period, about 2.5 million years ago. Therefore, about two-thirds of their height developed during the Quaternary Period. This upward thrust is still continuing.

There are numerous Quaternary volcanoes in Japan, including Fuji Volcano, renowned for its conical shape. At 3,776 meters, it is also the tallest volcano in Japan.

Many of the country's 200 volcanoes are still active, and some have been the cause of great disasters.

Massive earth movements have resulted in a variety of types of land collapse, including landslides, which are capable of destroying mountain villages.

As well, torrential rainfall can result in land collapse, due to the country's geological and geomorphic features. The archipelago is situated in the northeastern part of the Asian monsoon area, which features intense rainfall and abundant levels of precipitation. The Bai-u front, a short rainy season that occurs in June and July, and seasonal typhoons result in great amounts of rainfall.

According to official statistics, the maximum rainfall in Japan is 55.9 millimeters in 10 minutes, 87.9 millimeters in 30 minutes, 157.0 millimeters in an hour, 415.3 millimeters in 6 hours, 844.5 millimeters in 10 hours, 1,109 millimeters in 24 hours, 3,462 millimeters in one month and 10,216 millimeters in a year.

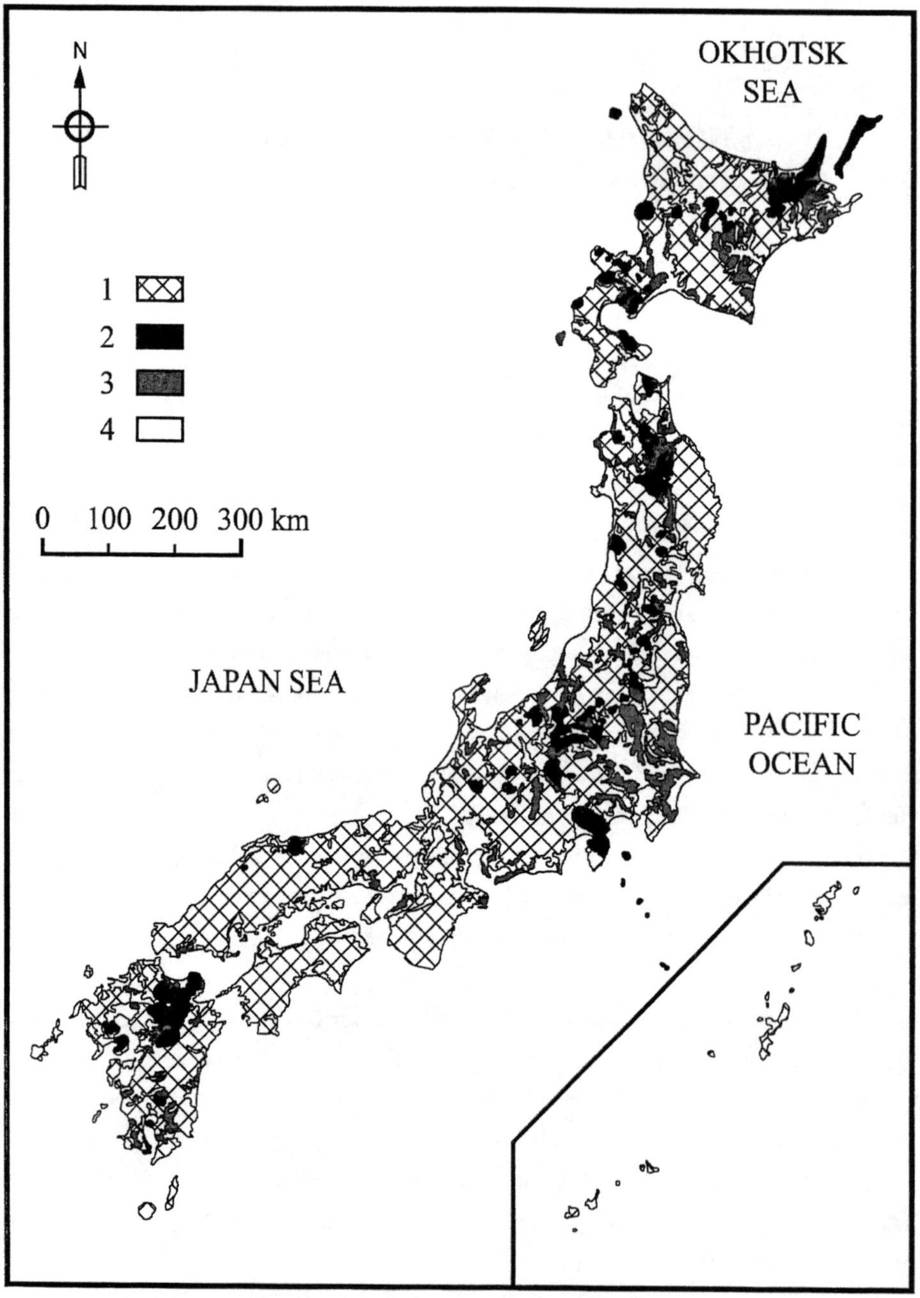

Figure 7.1 Geomorphological land classification of Japan
(1) Mountain region, (2) Volcanic region, (3) Terrace, (4) Alluvial plain.

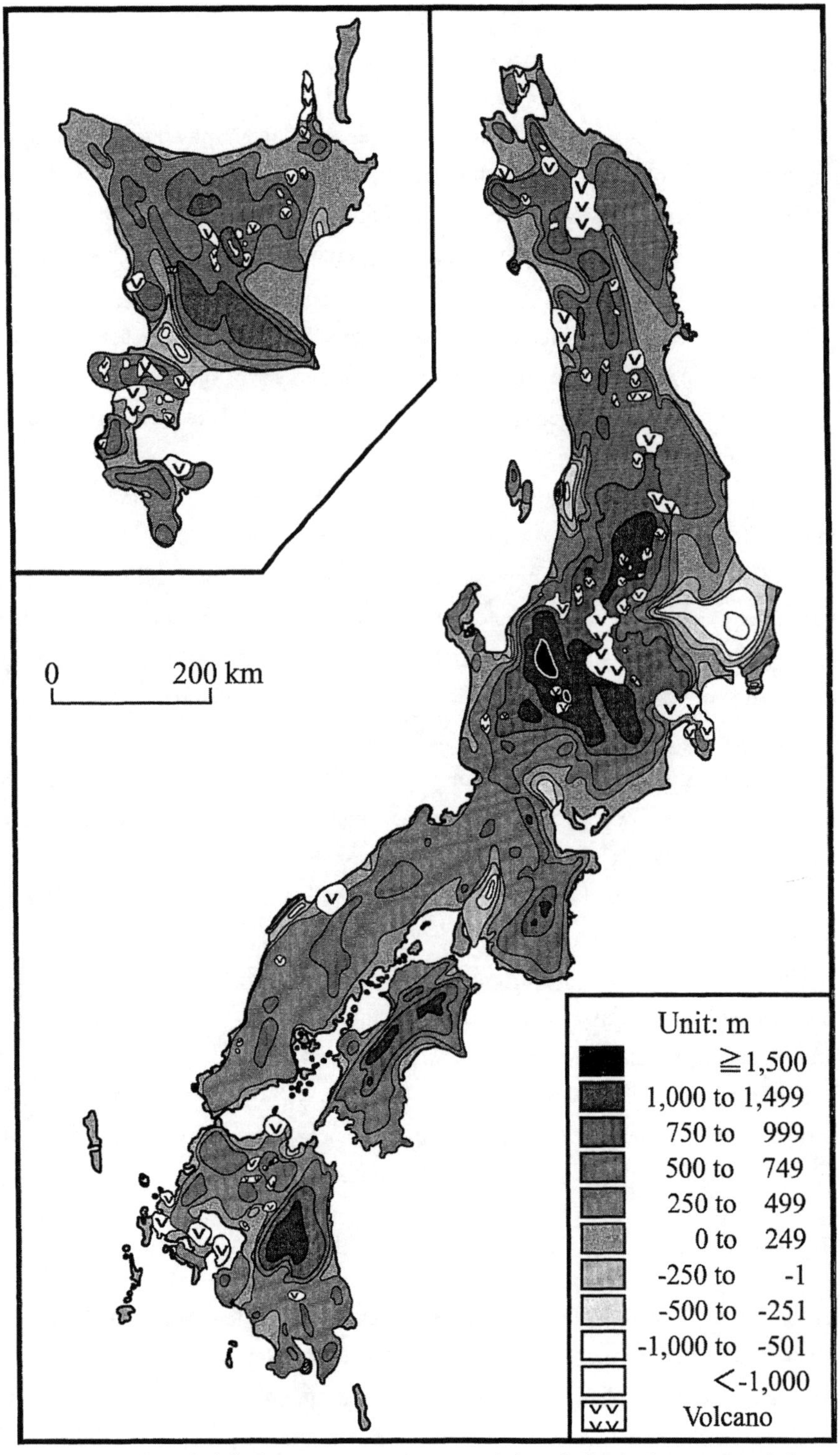

Figure 7.2 Vertical displacement of Japan's geomorphology during the Quaternary Period (National Research Center for Disaster Prevention, 1969; modified)

7.1.2. FEATURES OF THE KISO MOUNTAIN RANGE

The author wishes to discuss the relationship between topography and land collapse by using the example of the Kiso Mountain Range.

About 53.5 percent of the drainage area of the Kiso River basin is above 700 meters. There are many sharp slopes; more than 52.2 percent of the drainage basin is steeper than 30 degrees.

The relief energy of the area is high, with a maximum of 1,300 meters (*Figure 7.3*). Relief energies of over 700 meters comprise 30.8 percent of the area while those over 500 meters occupy 64.2 percent (*Table 7.1*). *Figure 7.4* shows the relationship between summit height and relief energy.

The drainage area is at an early mature or fully mature stage in the erosion cycle, and erosion continues.

The Kiso Mountain Range consists of a combination of gentle and steep slopes. The gentle slopes, surrounded by the steep slopes, are found on the top of mountains and mountain flanks. The range was formed by intermittent ground upheaval. The gentle slopes were formed during stable periods, and the steep slopes resulted from the river erosion that occurred during periods of upheaval.

The gentle slopes were formed during the Pliocene Epoch, more than 2.5 million years ago, and the steep slopes during the Quaternary Period. Landslides generally occur at the boundary of these steep and gentle slopes.

Upheaval continues. The region has risen between 5 to 11 centimeters during the last 50 to 60 years.

The area consists of granite, quartzporphyry, andesite, basalt, conglomerate, sandstone and slate. Generally speaking, the granite is easily weathered and the quartz-porphyry, andesite, basalt and sandstone are readily fractured.

Furthermore, there are usually more than two types of rock lying on top of one another, which increases the risk of collapse. For example, white Mesozoic rhyolite is covered by black Quaternary andesite in the Ontake Volcano.

7.1.3. LAND COLLAPSE IN THE NAKATSU RIVER BASIN

Because of the above-mentioned geomorphic and geologic features of the basin, as well as other basic factors such as vegetation, landslides follow heavy rainfall and a lot of sand and gravel is supplied to the river bottom. Rainfall in the upper reaches of the Kiso

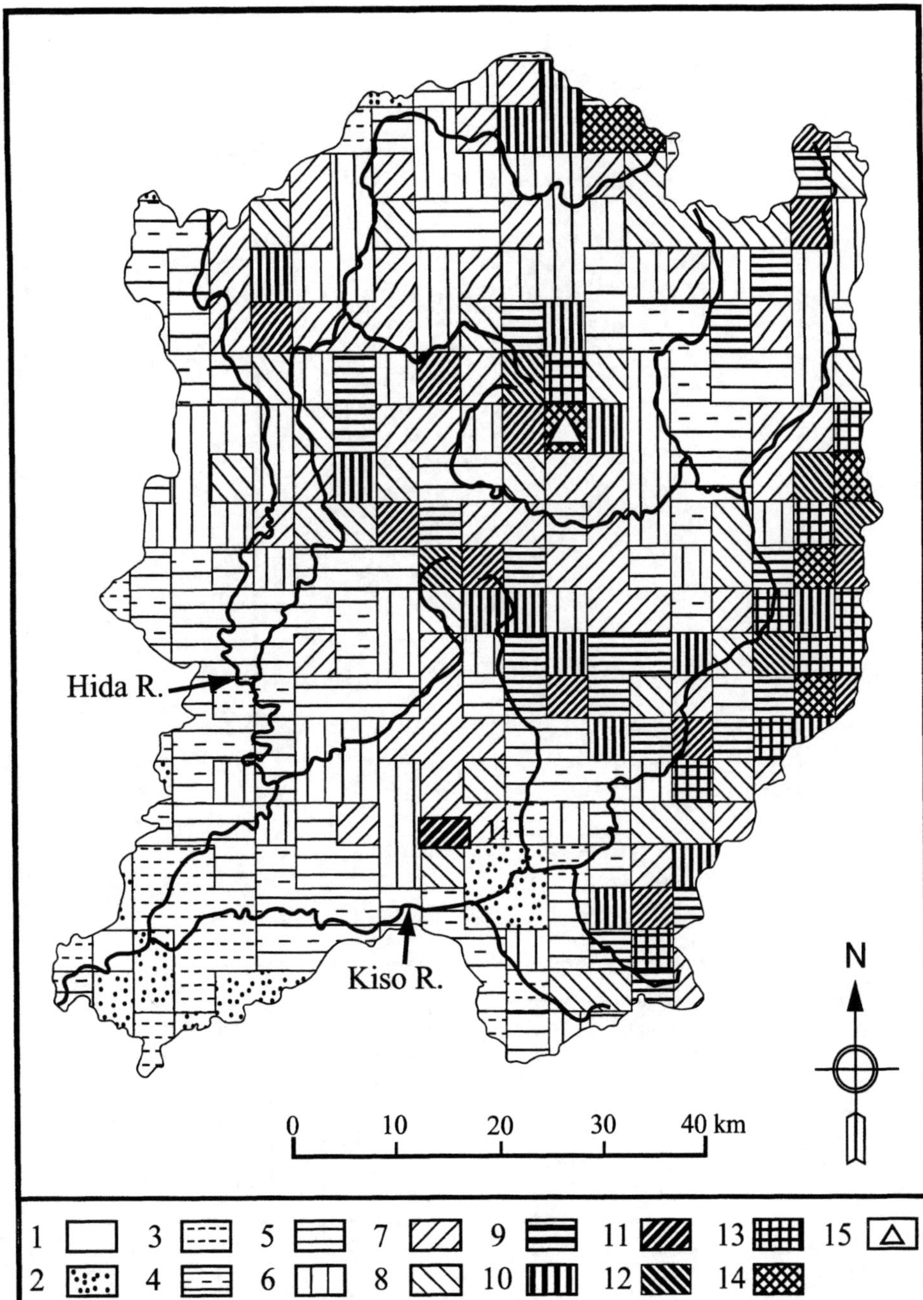

Figure 7.3 Relief energy of the Kiso River basin (Oya, 1973; modified)
(1) 0 m, (2) 100 m, (3) 200 m, (4) 300 m, (5) 400 m, (6) 500 m, (7) 600 m, (8) 700 m, (9) 800 m, (10) 900 m, (11) 1,000 m, (12) 1,100 m, (13) 1,200 m, (14) 1,300 m, (15) Ontake Volcano.

Rerief Energy	km²	%
0 m	15.92	0.3
100 m	164.32	3.3
200 m	253.12	5.2
300 m	532.64	10.8
400 m	790.88	16.2
500 m	896.96	18.3
600 m	742.96	15.1
700 m	491.92	10.0
800 m	307.04	6.3
900 m	237.60	4.8
1,000 m	168.40	3.4
1,100 m	77.20	1.6
1,200 m	143.68	2.9
1,300 m	87.84	1.8

Table 7.1 Rerief Energy of the Kiso River (Oya, 1973a; modified)

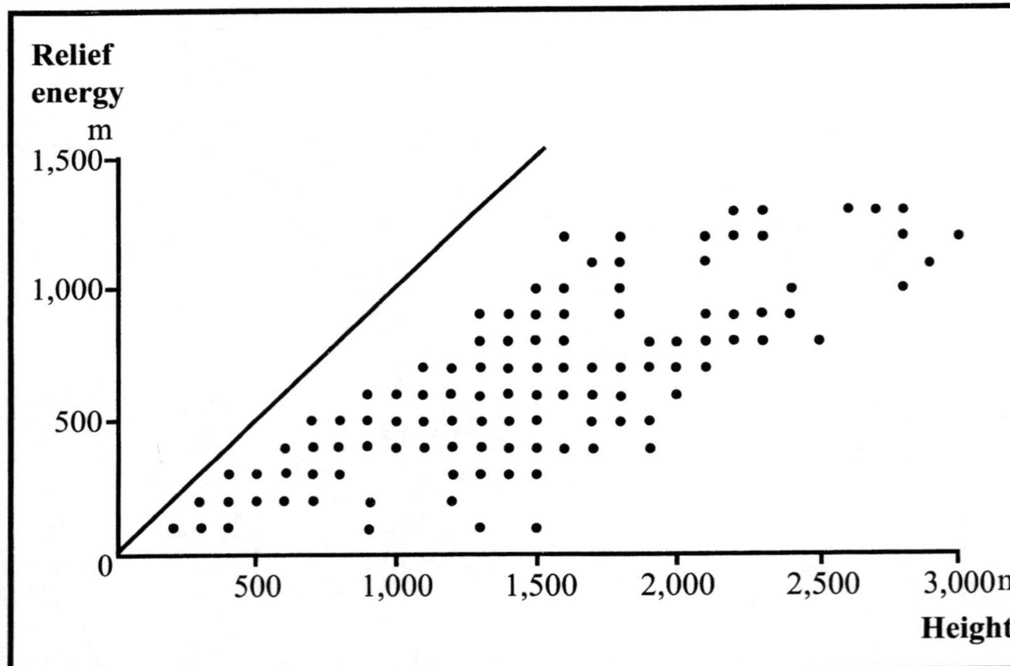

Figure 7.4 Relationship between height of summit and relief energy of the Kiso River basin (Oya, 1973a; modified)

averages more than 3,000 millimeters per year.

The knowledge that land collapse will likely occur at the boundary between gentle and steep slopes, both on mountain ridges and flanks, helps us predict the location and scale of damage following torrential rainfall.

A larger scale is needed to create geomorphological land classification maps for mountainous regions than for maps of the plains. The author has learned from experience that a scale of 1:20,000 serves this purpose, thus requiring that the scale for photographs be at least 1:20,000.

7.1.4. RELATIONSHIP BETWEEN GEOMORPHOLOGY AND LAND COLLAPSE

As shown by the map in the color section, almost all land collapse occurs at the point where a convex gentle slope, a mountain ridge, meets a concave steep slope along the valley. The convex slopes are reduced by land collapse.

When the author prepared a map showing the distribution of land collapse sites from aerial photographs taken in 1947, he found 139 such sites in the area. But in a map created by using aerial photographs taken in 1965, the number of land collapse sites had jumped to 2,165. The author attributes the dramatic increase to the effects of the Ise Bay Typhoon of 1959.

In general, there are many landslides in areas where convex slopes are narrow. In the Ataka-dani River basin area of the Kiso Mountains, for example, landslides are small in size but large in number. In contrast, there are few landslides in areas where convex slopes are wider.

There have been numerous incidents of land collapse in the Ochiai River basin, caused by erosion of the riverbanks.

7.2. Landslides from Volcanoes

7.2.1. FEATURES OF VOLCANOES IN JAPAN

Volcanic regions make up about a tenth of the country, and around 60 volcanoes are currently active.

Asama Volcano, in the central part of Japan, erupted in 1783. It first spewed volcanic ash, which was then followed by a pyroclastic flow and an eruption of lava. The so-called Agatsuma pyroclastic flow occurred on July 7, only to be followed by the so-called Kambara pyroclastic flow the next day. The latter flow devastated the village of Kambara, burying 400 people. Only 91 villagers managed to escape to a small temple situated on a higher level.

The pyroclastic flow reached the Agatsuma River in the upper reaches of the Tone River and jammed it. The dam broke, and the resulting flood caused great damage along the river course. About 1,050 people and 500 horses drowned, and 1,061 houses were washed away.

Historical records tell us of the many eruptions and disasters that have occurred in the volcanic regions, which are among the most scenic areas in Japan. Calderas have frequently become volcanic lakes, and there are numerous hot springs in these areas, which are now the base of a great many national parks.

In this chapter we shall discuss disasters caused by land collapse or erosion by torrential rainfall or flooding in these volcanic regions.

7.2.2. LANDSLIDES AND FLOODING IN THE TARADAKE VOLCANO AREA

The Taradake Volcano is east of Nagasaki, a city known for its close relationship with the Dutch during the Edo Period. South and southwest of the volcano are the small cities of Isahaya and Omura.

Classified as extinct, the volcano is composed of alternate layers of strata, including lava, pyroclastic flow and andesite. Its base, visible in the southern part of the foot, was formed during the Tertiary Period and has been flattened by erosion. The surface is covered by a plateau of lava. At present the volcano is dissected by the Honmyo River, its branch and many streams.

The area was devastated by a disastrous rainfall and flood on July 25, 1957. A total of 1,100 millimeters of rain fell in one night, and more than 600 people lost their lives. The disaster was called the Isahaya Suigai (Flood).

Immediately after the flood, the author and Yoshimi Ichinose used field surveys and aerial photos to make a map showing the state of flooding and land collapse around Isahaya and the Taradake Volcano (*Figure 7.5*). A lot of the land collapse occurred on the southern slopes of the Taradake Volcano, and was characterized by the following:

a) The landslides occurred in the middle and lower parts of the Taradake Volcano, which are mainly made up of hypersthene andesite and pyroclastic flow.

b) Each landslide was shaped like a tadpole, with the jet point at its head. In general, the eroded material formed a thin layer.

c) Water that had infiltrated into the mountain slopes and then burst out was responsible

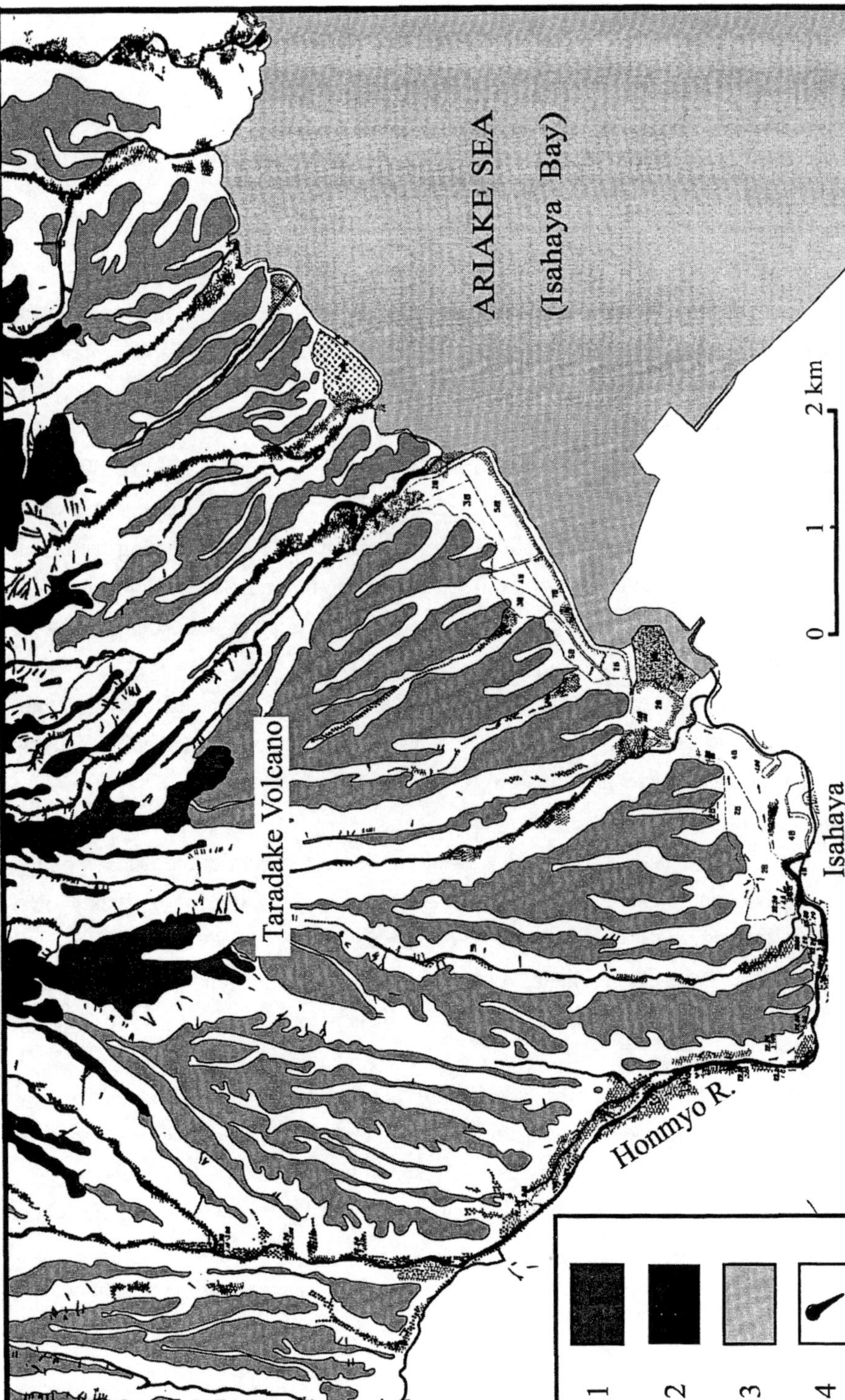

Figure 7.5 Geomorphological survey map of the Isahaya Bay area (Oya, 1959; simplified) (1) Hypersthene andesite, (2) Pyroclastic flow, (3) Sea, (4) Landslide.

for the majority of the landslides.

d) While some landslides occurred in the hollow valley slopes, where water collects quickly, most occurred on flat slopes that do not retain water.

e) The landslides resulted from the changes in physical properties that occurred when water traveled between the hypersthene andesite and pyroclastic flow (Tada, Oya and Ichinose, 1959).

The following flood types were seen:

a) Flooding in the valleys situated radially at the foot of the volcano.

A lot of the landslides occurred in the upper courses of the valleys, which consist of hypersthene andesite, lava and pyroclastic flow, and are situated radially around the extinct crater of the Taradake Volcano.

b) Flooding in the plains surrounding the foot of the volcano.

Here, the Honmyo River has tributaries from the northeast, but none from the west nor south. The velocity of the flood waves in these valleys was very fast, and the water reached high levels throughout the plains. This was caused by the following factors: It had rained heavily along the valley, the floodwater from the radial valley plains flowed into the main course at about the same time, and no stagnation could occur because of the steepness of the valley plain slopes.

Floodwater destroyed the banks in many places, and streams changed their meandering course to take more direct routes. The water removed surface soil, exposing the layer of sand and gravel at points.

c) Flooding in the lower part of the canyon.

The Honmyo River cuts the foot of the volcano in the western part of the Isahaya region to form a small canyon. At the lower part of the canyon, where the city of Isahaya lies, floodwater flowed radially to the east. Gravel from the canyon, banks and riverbed were deposited at the areas where banks were destroyed, and sand and silt accumulated beyond these deposits. The floodwater was 3 to 4 meters deep.

d) Flooding in the delta.

In the delta east of Isahaya, the velocity of the floodwater was slow and it was submerged for a long period. Most of the delta was underwater from two to seven days, but parts were submerged for as long as 30 days. The water was about 2 meters deep.

The area where the coast banks had collapsed—not by tidal waves but by inland water—suffered salt damage.

Deposited silt was rarely seen along the river course, and greater damage was caused by stagnation.

CHAPTER 8

JAPAN'S COASTAL LAKES

Japan's many lakes can be divided into three groups: those formed by volcanic activity, including Toya and Shikotsu lakes in Hokkaido; those formed by crustal movement, including Biwa Lake—the country's largest lake—in western Japan and Suwa Lake in the central part of Japan; and coastal lakes formed by marine and fluvial action, including Kasumigaura and Kitaura lakes in the central part of Japan.

The author wishes to discuss the geomorphic features of Japan's lakes and how they are affected by human activities, using Kasumigaura Lake and northern Honshu's Ogawara Lake as examples.

8.1. Geomorphic Features of Coastal Lakes

Lacustrine lowlands, with heights of 0 to 5 meters and widths of 200 to 700 meters, can generally be found along the shorelines of Japan's coastal lakes (*Figure 8.1*). There are also littoral shelves, which are only a few meters deep (Hirai, 1989).

The lacustrine lowlands are made up of two Holocene Terraces, and have beaches or littoral marshes. The higher terrace (Lacustrine Terrace I) is 2 to 5 meters in height and 100 to 200 meters in width, while the lower one (Lacustrine Terrace II) is 1 to 2 meters in height and 100 to 500 meters in width. The two terraces were formed at different lake levels and are connected with the high sea levels of the Holocene. It is assumed that Lacustrine Terrace I was formed about 6,000 years B.P., and Lacustrine Terrace II was formed about 4,000 to 3,000 years B.P. (Hirai, 1983; Oya et al., 1989).

The littoral shelf can be divided into two surfaces based on the differences in depth, sediment and shape. Littoral Shelf I develops widely and continuously at a depth of 0.5 to 2 meters, and a width of 200 to 300 meters. Its surface is now in the process of forming at the present water level. Littoral Shelf II forms as the topset bed of a river delta, a tidal delta, a sand spit or a wave cut bench, at a low water level. It is assumed that low water levels can be correlated to low sea levels, which occurred three times after post-glacial transgressions; that is, about 4,500 years B.P., 3,000 years B.P. and in

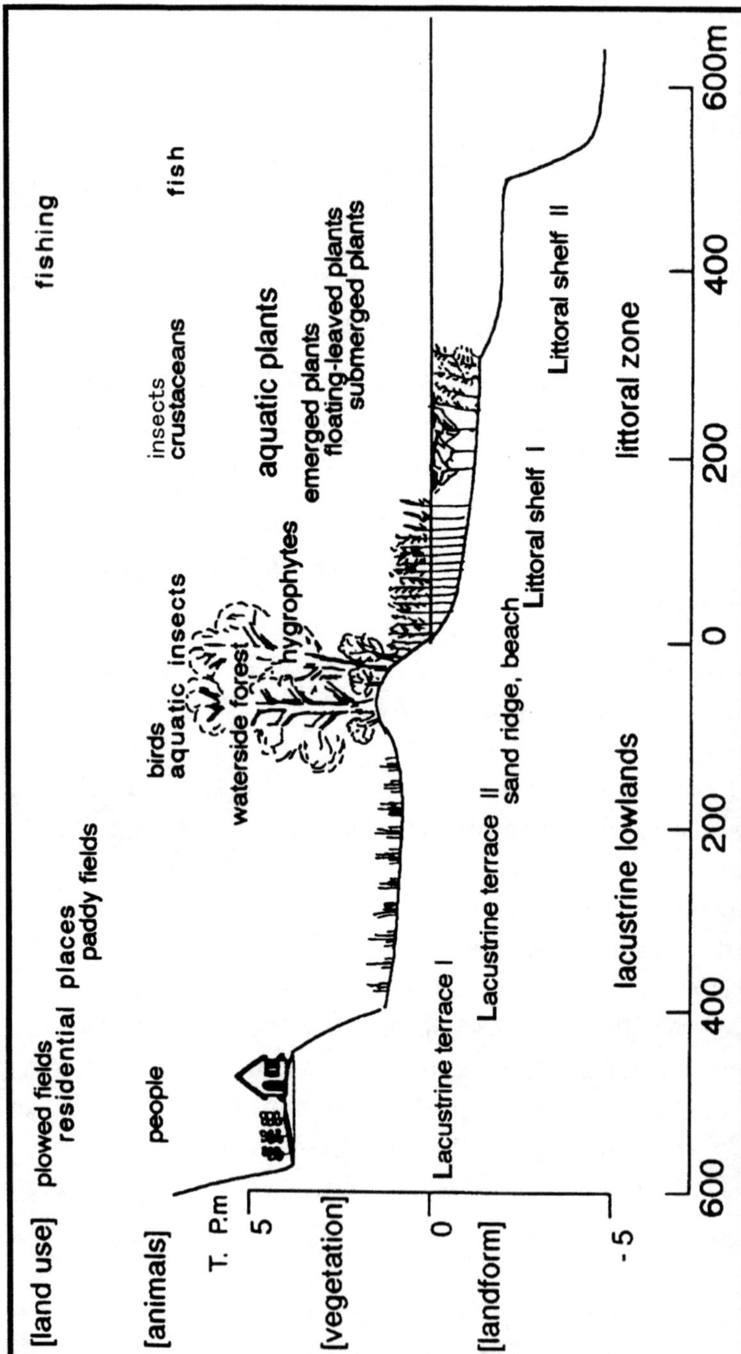

Figure 8.1 Schematic natural landscape and traditional land uses of a coastal lakefront (Hirai, 1992; modified)

the 16th to 17th centuries. Beaches and Littoral Shelf I tend to form around Lacustrine Terrace II. (Hirai, 1987, 1989; Oya et al., 1994).

8.2. Method of Coastal Lake Map Preparation

Using aerial photographs, the target area is first classified by major geomorphic elements such as terraces, valley plains, natural levees, back-marshes, former river courses, deltas, sandbanks, sand spits, sand dunes and wave cut benches. The results of field surveys of the area are then checked.

In preparing the sublacustrine map of a lake, its bottom is first classified by major geomorphological elements such as littoral shelves, sublacustrine valleys and sub-lacustrine plains, using an iso-depth chart created by the Geographical Survey Institute of the Ministry of Construction. Secondly, the depth of the lake is measured with a Fathometer.

8.3. Kasumigaura Lake

Kasumigaura and Kitaura lakes are located in the eastern part of the Kanto Plain, northeast of Tokyo (*Figure 1.2*). With a size of 171 square kilometers, Kasumigaura Lake is the second largest lake in Japan and the largest to be formed by marine and fluvial action.

Even the deepest parts of the two lakes are quite shallow, only 8 meters, and the average depth is only 4 meters. Many small rivers, i.e. the Sakura River, flow into both lakes. The total drainage basin of these rivers is 2,200 square kilometers.

These lakes are connected to the Pacific Ocean by the Tone River. The water of the lakes was brackish before the construction of a dam in the lower reaches of the Tone River.

8.3.1. FEATURES OF THE KASUMIGAURA LAKE AREA

The area consists of diluvial terraces and alluvial plains (*Figure 8.2*). The diluvial terraces are separated by Kasumigaura Lake, Kitaura Lake and the Tone River. Tectonic movement has caused the ground to become higher as it nears the Pacific Ocean.

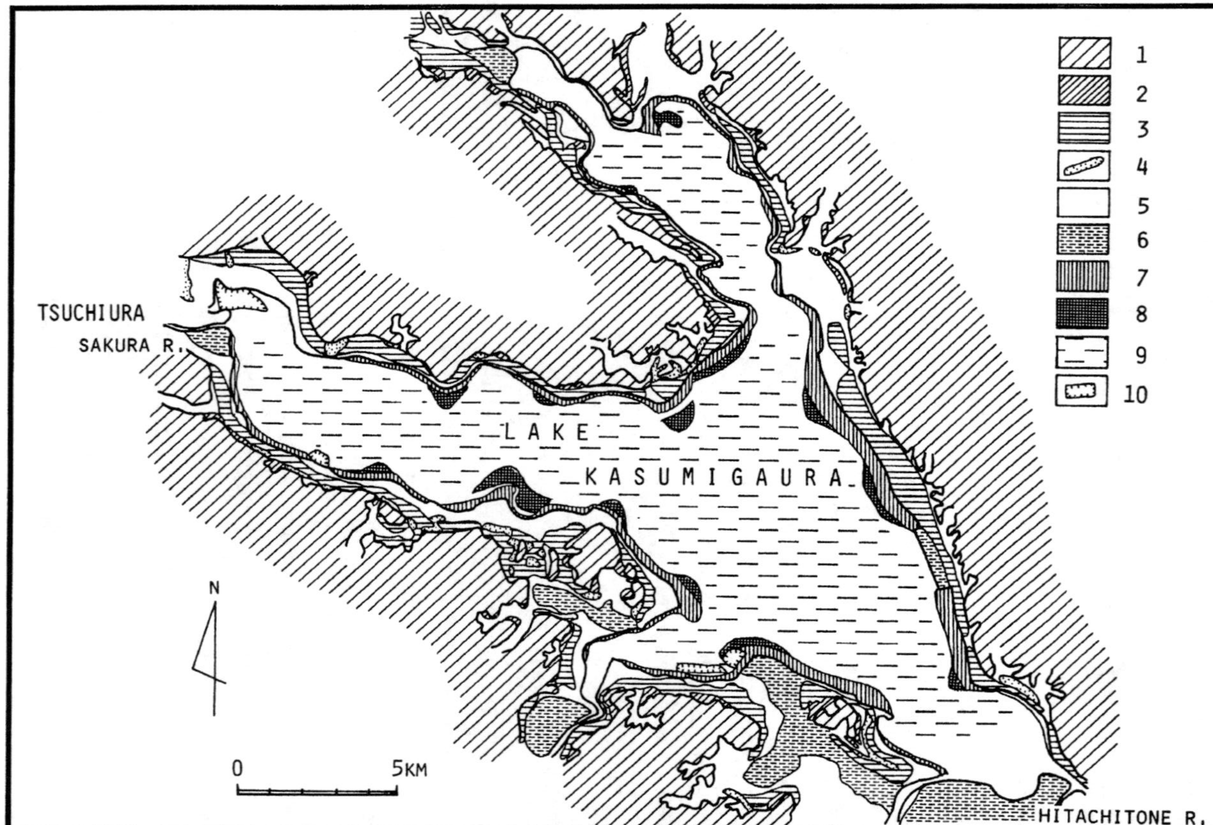

Figure 8.2 Geomorphological land classification map of Lake Kasumigaura (Oya, Haruyama and Hirai, 1989)
(1) Diluvial terrace, (2) Lacustrine Terrace I, (3) Lacustrine Terrace II, (4) Elevated sandbank or sand spit, (5) Alluvial lowland, (6) Reclaimed land, (7) Littoral Shelf I, (8) Littoral Shelf II, (9) Sublacustrine plain, (10) Artificial trench.

The diluvial terraces are divided into upper and lower terraces. The former was formed about 120,000 to 130,000 years B.P. (Last Interglacial Age) and the latter, 10,000 to 30,000 years B.P. (Last Glacial Age).

These terraces are surrounded by cliffs and dissected by small streams.

Alluvial terraces from the Holocene formed at the foot of the diluvial terraces. These terraces are divided into two groups: upper alluvial terrace (Lacustrine Terrace I) and lower alluvial terrace (Lacustrine Terrace II).

Lacustrine Terrace I is about 4 to 8 meters high, and Lacustrine Terrace II, about 1.8 to 3 meters high. The former was formed during the Holocene Transgression (Jomon Transgression), between 6,000 to 4,000 years B.P. (*appended Table 3*), and on it are sand spits, ancient tombs and shell mounds. From the air, one can see how the area is marked into square-like sections—the remains of ancient rice field grids.

Lacustrine lowlands have developed along Kasumigaura Lake and other small lakes, and land has been successfully reclaimed between the southern part of Kasumigaura Lake, Kitaura Lake and the Tone River.

The alluvial plain along the Tone River consists of valley plains, sand spits or sandbanks, natural levees, back-marshes, deltas, abandoned river courses, reclaimed land, etc.

The sand spits or sandbanks, formed at the foot of the terraces, are divided into several groups: Sand Spits I, II and III; and Sandbanks I, II and III. Several ancient tombs have been discovered on the surface of Sand Spit I. Both Sand Spit I and Sandbank I were formed during the Holocene Transgression.

There are few natural levees in the area. Back-marshes or deltas fill the spaces between the terraces and natural levees, sandbanks and rivers.

8.3.2. TOPOGRAPHY OF THE KASUMIGAURA LAKE BED

Littoral shelves, with widths of 200 to 500 meters, are well developed along the lakeside. The shelves are divided into two groups: Littoral Shelf I (-1.0 to -1.5 meters) and II (-2.0 to -3.5 meters). Small cliffs or steep slopes are found between the two shelves.

Large sublacustrine plains are also well developed in the Kasumigaura and Kitaura lakes. Steep slopes lie between the sublacustrine plains and littoral shelves.

Sublacustrine valleys, believed to have been formed by erosion due to littoral currents, are found in several places.

8.3.3. GEOMORPHIC DEVELOPMENT AROUND KASUMIGAURA AND KITAURA LAKES

Kasumigaura and Kitaura lakes are connected with the Pacific Ocean via the Tone River. Therefore, the geomorphology of the two lakes and their vicinity is influenced by eustatic movements.

During the Würm Ice Age, the Kinu River dissected a large valley lying under the Kasumigaura Lake and the lower reaches of the Tone River. During the Jomon Transgression, Kasumigaura, Kitaura and other lakes, as well as the lower reaches of the Tone River, were covered by seawater to form a large inland sea—the Katori Sea. Many people lived along its coast, producing a great number of shell mounds there.

The sea gradually disappeared due to the lowering of water levels, the deposition of sand and clay by littoral or fluvial actions, and human activities.

First to emerge were the sand spits and lacustrine terraces, and people moved there from the diluvial terraces.

Since the Meiji Restoration, the lower reaches of the Tone River have changed significantly due to river improvement projects.

8.4. Ogawara Lake

Ogawara Lake is situated at the root of the Shimokita Peninsula in northern Honshu.

With an average depth of 11 meters and 25 meters at its deepest, the brackish lake has a drainage basin of 65.6 square kilometers. The Shichinohe River and other small rivers flow into the lake, which connects to the Pacific Ocean via the Takase River (*Figure 8.3*). The total drainage basin is 866.9 square kilometers.

8.4.1. FEATURES OF THE OGAWARA LAKE AREA

The terraces in the area are divided into three groups: Shichihyaku, Furumaki and Misawa. Misawa Terrace can be further divided into high, middle and low terraces. Peri-glaciation formed wavy surfaces on these terraces.

The lacustrine terrace is well developed around the lake. The author believes Lacustrine Terrace I (3 to 4 meters above sea level) was formed by deposition and Lacustrine Terrace II (1 to 1.5 meters above sea level) was formed by erosion.

The plain located along the lower course of the Shichinohe River is the largest in the area. The combination of the geomorphic elements of the plain is as follows:

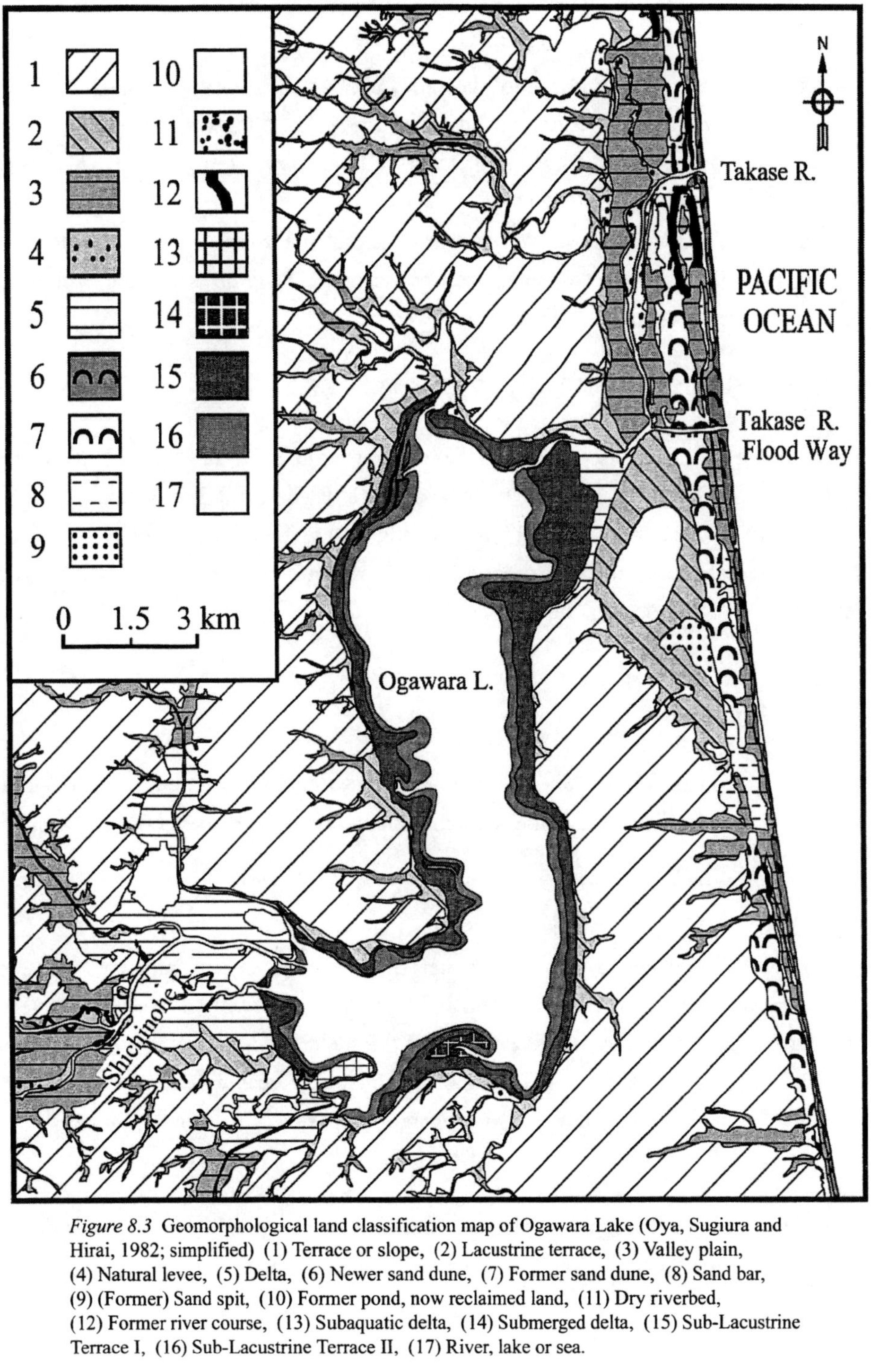

Figure 8.3 Geomorphological land classification map of Ogawara Lake (Oya, Sugiura and Hirai, 1982; simplified) (1) Terrace or slope, (2) Lacustrine terrace, (3) Valley plain, (4) Natural levee, (5) Delta, (6) Newer sand dune, (7) Former sand dune, (8) Sand bar, (9) (Former) Sand spit, (10) Former pond, now reclaimed land, (11) Dry riverbed, (12) Former river course, (13) Subaquatic delta, (14) Submerged delta, (15) Sub-Lacustrine Terrace I, (16) Sub-Lacustrine Terrace II, (17) River, lake or sea.

Valley Plain + Delta + Subaquatic Delta

A stream dissects the upper part of the valley plain. Natural levees are small.

The height of the main section of the delta is 1 to 2 meters, with some parts below sea level, and it was covered by seawater during the Jomon Transgression. The lower part of the delta is now reclaimed land.

Transversal sand dunes are well developed along the coast of the Pacific Ocean. They can be classified into two categories: newer dunes, with no vegetation, and older dunes, which are covered with vegetation.

8.4.2. TOPOGRAPHY OF THE OGAWARA LAKE BED

The author has divided the littoral shelf into three groups: Littoral Shelf I (0 to -0.5 meters), Littoral Shelf II (-1.5 to -2 meters) and Littoral Shelf III (-3 to -4 meters).

The lake delta has been divided into three kinds: tidal delta, located near the Takase River; subaquatic delta, located near the mouth of Shichinohe River; and submerged delta, located in the southern part of the lake.

8.5. Map Utilization

Development plans for the lake focus on water use and flood control. Furthermore, conservation of the lake's natural environment has recently become the object of much public attention.

Water Use

According to these plans, the lake's water level will be slightly raised or lowered depending on circumstances. From April to June, the water will be raised 0.8 meters, and during the dry season, it will be lowered 0.8 meters. Knowing the distribution of lacustrine terraces and shelves helps us to predict which areas will be inundated when the water is raised and which will dry up when it is lowered.

Flooding

The features of inundation in the plains around Kasumigaura and Ogawara lakes can be predicted by use of a geomorphological land classification map.

Lacustrine Terrace I, whose relative height above the present water level is more than 2 meters, is never subjected to any flooding, while Lacustrine Terrace II, whose relative height is less than 2 meters, will be partially inundated by extraordinary flooding.

Kasumigaura Lake flooding occurs under two circumstances: 1) The lake is connected with the Tone River through the Hitachi-Tone River, but the capacity of the Hitachi-Tone River to hold water is limited. Thus, when the level of the Tone River rises, the water intrudes into Kasumigaura Lake through the Hitachi-Tone River. 2) When it rains heavily in the Sakura River area, water pours into Kasumigaura Lake.

The first type of flooding occurred in 1938, pushing up the water level to 3.34 meters above sea level and causing serious damage to the lakeside lowlands, i.e. lower alluvial terraces, lacustrine lowlands, etc. The second type of flooding occurred in 1941.

Rising sea levels

Recently, the issue of global warning has received much attention. If the Earth's temperature rises, so will sea levels. Recent studies predict that the sea will rise 50 centimeters within the next 100 years. Many kinds of disasters will follow, including agricultural damage caused by the flooding of paddies and upland crop fields in the lower part of the plain and the salifying of soil. As well, tidal waves resulting from typhoons or tsunami from earthquakes will increase in number, bringing much tragedy. Geomorphological land classification maps are useful in predicting which areas will be inundated.

Conservation of the natural environment

In littoral marshes and on Littoral Shelf I, plant communities are found widely at altitudes of less than 0.5 meters and depths of less than 0.5 meters. These aquatic plants play important roles in the ecosystem of coastal lakes. The littoral zones create a spawning and breeding ground for fish, crustaceans and other marine life. And such a littoral zone improves the water quality.

Since the end of World War II, a great many coastal lake regions have been reclaimed for agricultural purposes. Recently, land has been frequently reclaimed for housing developments and urban infrastructure. It is becoming increasingly rare to find natural shorelines, free from artificial structures.

The Kasumigaura Lake region's littoral zone has been transformed by several types of urban development. Aquatic plants, so important to the ecosystem of coastal lakes and to improving water quality, have all but disappeared. And the lakefront scenery has been destroyed.

In other coastal lakes, including Ogawara, some significant environmental changes occur when a continuous littoral bank and a closing levee at the inlet of lake are constructed, or when the water level is artificially increased to supply the growing demands of neighboring cities.

More attention must be paid to environmental changes in and around the coastal lakes. Greater efforts are needed to preserve the ecosystem and protect it from the onslaught of urbanization. Geomorphological land classification maps are useful in that regard.

CHAPTER 9

FLOOD CONTROL IN TOKYO, OSAKA AND NAGOYA

9.1. Flooding in the Tokyo area

9.1.1. OUTLINE OF TOKYO'S GEOMORPHOLOGY

Tokyo and the national capital region are located in the Kanto Plain, the largest plain in Japan. The region's topography consists of alluvial plains, diluvial terraces and dissected small valleys in the terraces. The diluvial upland has been free from flooding, with the exception of the flood caused by the Kano River Typhoon in 1958. However, the alluvial plains have been inundated numerous times.

Tokyo's lowlands consist of Katsushika, Edogawa, Sumida, Koto and Adachi wards. The plain was formed by the deposition of the Tone River. At present, the river pours directly into the Pacific Ocean, but in former days it poured into Tokyo Bay. Construction was carried out from 1594 to 1809 to divert the Tone River into the lower reaches of the Kinu River to flow directly into the Pacific Ocean.

The former course of the Tone River is now called the Naka River. Sand bars, natural levees, back-marshes, old lagoons, former river courses, deltas, reclaimed land and artificially filled fields are all important topographical elements in the Naka River basin. The natural levees formed by the Furu-Tone River—the main course of the Tone River during the Edo Period—are significant. The land becomes higher as we approach the river. When flooding occurs, the water overflows the banks and floods the adjacent area. Sand bars were formed along the former coastline, which shows the fluctuation of sea levels. The delta is divided into two parts: upper and lower. The upper delta ranges from 2 to 5 meters above sea level (*Figure 1.3*).

9.1.2. TYPHOON CATHERINE

On Sept. 15, 1947, the Tone River basin was battered by Typhoon Catherine. The maximum flood discharge along the Tone River was estimated to be 15,000 cubic meters per second at Yattajima Research Point. The embankment of the Tone River broke near the town of Kurihashi. The floodwater flowed along the ancient Tone River courses, reaching Tokyo three days later. Water from the floods of 1704, 1785, 1846, 1907 and 1910 also reached Tokyo, and it is likely that this pattern will be repeated in future floods.

The flood's isochronal lines were at close intervals except where embankments were breached. The intervals were wide in marshy lands, where the flood current was slower.

The Naka River basin is an urbanized area, and man-made structures such as buildings, railway lines and roads proved larger obstacles to the flow of the current than the area's natural topography. While the flooding was not very deep, it inundated a large area and stagnated for a long period.

9.1.3. THE KANO RIVER TYPHOON

The Kano River Typhoon pounded Tokyo from Sept. 25 to 27, 1958. The maximum daily precipitation reached 261 millimeters, but at the time, the drainage capacity was only 40 millimeters per hour and even the uplands, normally unaffected by flooding, were inundated.

In the lowlands, the water stagnated for as much as eight days, partly because of the low gradient of the land, partly because of the narrowness of the rivers and partly because the water level in the river was higher than the height of the ground in protected areas. Tidal differences also affected the drainage.

In the Ooba River system, the water levels in the river and the protected area were nearly the same. The depth of the stagnant water was 25 to 130 centimeters in Katsushika Ward and 40 to 120 centimeters in Edogawa Ward, and stagnation lasted as long as one week.

In the Naka River basin, the water level of the river was higher than the height of the protected areas, resulting in stagnant water with a depth of 50 to 150 centimeters (*Figure 9.1*).

9.1.4. RECENT FLOODING

Typhoons in 1976, 1979, 1981 and 1982, as well as localized torrential downpours in 1978, 1979 and 1981, caused extensive damage in the lower delta, especially in a former lagoon located below sea level between the Naka and Edo rivers behind elevated sand bars.

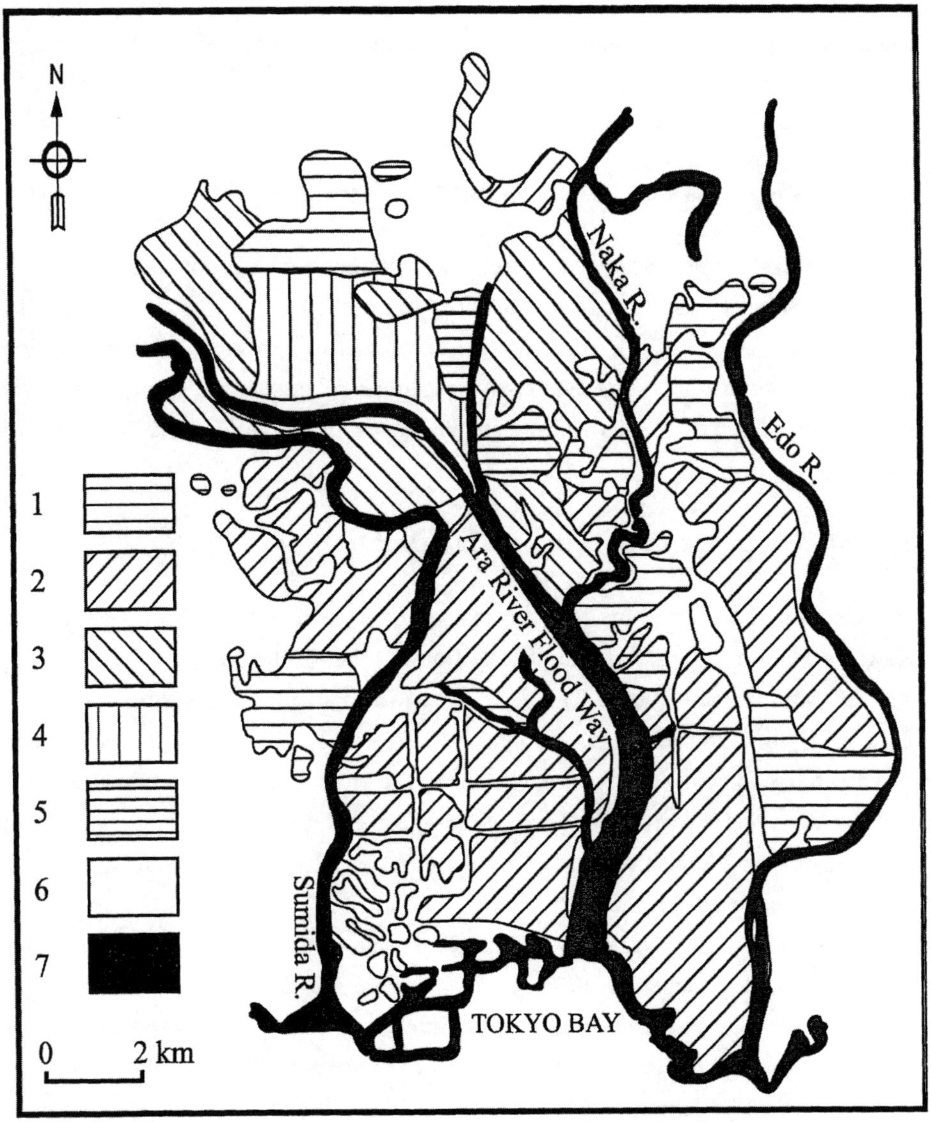

Figure 9.1 Area submerged by the 24th Typhoon of 1958 (Tokyo Metropolitan Office, 1958; modified) (1) Area submerged 0-50 cm, (2) Area submerged 50-100 cm, (3) Area submerged 100-150 cm, (4) Area submerged 150-200 cm, (5) Area submerged more than 200 cm, (6) Uninundated area, (7) River.

But in the last 10 years, flooding has been minimal compared to that which occurred during the Catherine and Kano River typhoons.

To control flooding, multipurpose dams have been constructed and embankments strengthened in the upper reaches of the Tone River. In addition, larger facilities for rainwater management have been constructed along the Naka River. Thanks to these public works projects, damage has slowly been reduced. When there is a large rainfall, however, inundation still occurs in these areas (Oya and Haruyama, 1987).

9.2. Inundation Caused by Rainfall in the Osaka Area

During the last several years, there has been significant inundation from rainfall in Japan's urban areas. Eastern Osaka is a typical example.

The nature and extent of inundation can be estimated by using a geomorphological survey map of the region that exhibits the relationship between the topography of the Neya River basin near Osaka and inundation caused by rainfall in July 1967 (*Figure 9.2*).

9.2.1. TOPOGRAPHY OF THE OSAKA PLAIN

The Osaka Plain, situated in the lower reaches of the Yodo River, is divided into the Kawachi Plain in the east and the Osaka Coastal Plain in the west by the Agemachi Terrace. Most of the Kawachi Plain is encompassed by the Neya River basin, which is bordered by the Ikoma Mountains in the east, Agemachi Terrace in the west, Yamato River in the south and Yodo River in the north. The ground rises toward the riverbanks.

The Yodo River was formerly bifurcated into two rivers. One ran along the present Yodo River course and the other along the Furukawa River, which runs through the cities of Neyagawa and Kadoma.

The Yamato River was formerly bifurcated into three river courses: the Nagase, Tamagushi and Hirano. In 1704, a river improvement project closed off these distributors and a new channel was created, which is now the present river course.

The natural levees along the three former rivers are large. In general they run south to north, but in the central part of the Kawachi Plain, they run east to west, or parallel with the present Neya River. This river's course appears to be the boundary between inundation from the Yodo and Yamato rivers.

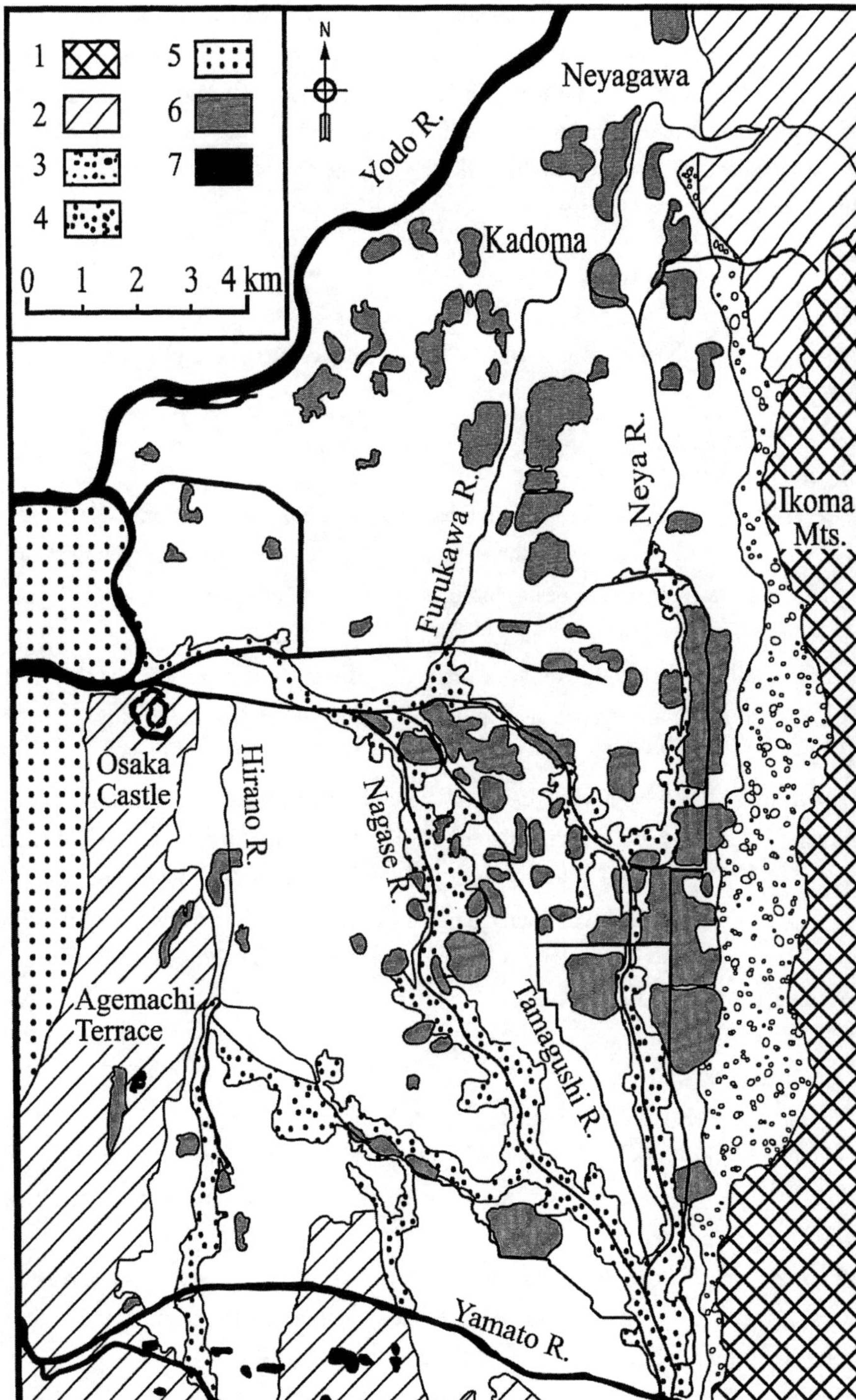

Figure 9.2 Relationship between the topography of the Neya River basin and the July 1967 inundation (Oya and Nakamura, 1969; modified)
(1) Mountain, (2) Terrace, (3) Fan, (4) Natural levee, (5) Sand spit, (6) Inundated area, (7) River.

In the Kawachi Plain, there are distinct regional differences in topography between the Yamato River basin and the Yodo River basin. The area formed by the Yodo River is lower than that of the Yamato River and has a gentler slope, indicating that the fluvial deposition of the Yodo River was weak in the past and that deposition will continue. The author surmises that floodwater will flow along the former river courses during future floods.

9.2.2. CAUSES OF INUNDATION

Inundation is a result of a) natural causes, b) man-made causes and c) a combination of a) and b).

a) Natural Causes

1) The basin of the Neya River is surrounded by hilly areas, impeding drainage.

2) Ground upheaval continues in the Agemachi Terrace and Ikoma Mountains, while the plains of the Neya River basin continue to lower. During the Holocene Age, the ground has been dropping from 1 to 1.9 millimeters per year.

As a result of river conservancy, overflowing from these rivers has halted, as has deposition of sand and gravel. The plain has thus lowered, making drainage more difficult.

3) The Yamato River basin contains many natural levees, which were formerly part of the main Yamato River and its distributaries. There are many back-marshes, which are surrounded by these natural levees, and inundation occurs easily in this area. The northern part of the Neya River drainage basin was formed by deposition from the Yodo River. The area is lower, the slope is gentler, and inundation occurs frequently because the Yodo River carries few deposits.

b) Man-made Causes

1) Urbanization has resulted in rapid rainwater runoff in recent years. Trees have been cut down in hilly areas, and irrigation ponds have all but disappeared in the plains. Drainage to the Neya River from the adjacent areas has become more difficult, and floods have been increasing.

2) When paddies are developed into urban areas, their irrigation canals are converted to sewers. As well as the fact that such canals are not suitable as sewers, they are usually in poor condition, as farmers often have done little to maintain them.

3) Roads are constructed without any consideration of drainage.

4) Urbanization has moved more quickly than the construction of sewers.

c) Combination of a) and b)
1) Ground lowering, another cause of inundation, results not only from tectonic ground movement but also from the overpumping of ground water. In the Neya River region, the ground fell 120 centimeters from 1935 to 1965.

The author stresses that we must consider human as well as physical factors when studying flood conditions (Oya and Nakamura, 1969).

9.3. Integrated Flood Control: Nagoya

9.3.1. NONSTRUCTURAL FLOOD CONTROL

Urbanization not only causes obstruction to natural drainage but also augments the rate of runoff and velocity. As a countermeasure to such a situation, it is necessary to introduce comprehensive flood controls. The concept was first initiated in the United States and Canada.

After World War II, a decades-long study of flood plain management by Gilbert White and his group of researchers gained attention as they pursued the problems of increased urban occupancy of the flood plain (Dougal, 1969). This concept emphasized nonstructural methods such as the use of building codes.

9.3.2. CITY PLANNING FOR THE MITIGATION OF FLOOD DAMAGE

Nagoya was the first municipality to implement nonstructural flood controls in Japan. Following the Ise Bay Typhoon in 1959, the city authorities of Nagoya and the Ministry of Construction created measures for comprehensive flood prevention. On the basis of the flood features and the damage that occurred, as well as topographic elements, the southern part of the city was divided into five zones and building regulations were issued for each zone (*Figures 9.3*).

All future buildings in the first zone, which included the artificially filled land around the port of Nagoya, were to be constructed in concrete. Buildings in the second area, in which reclaimed land was bordered by artificially filled land on the seaside and

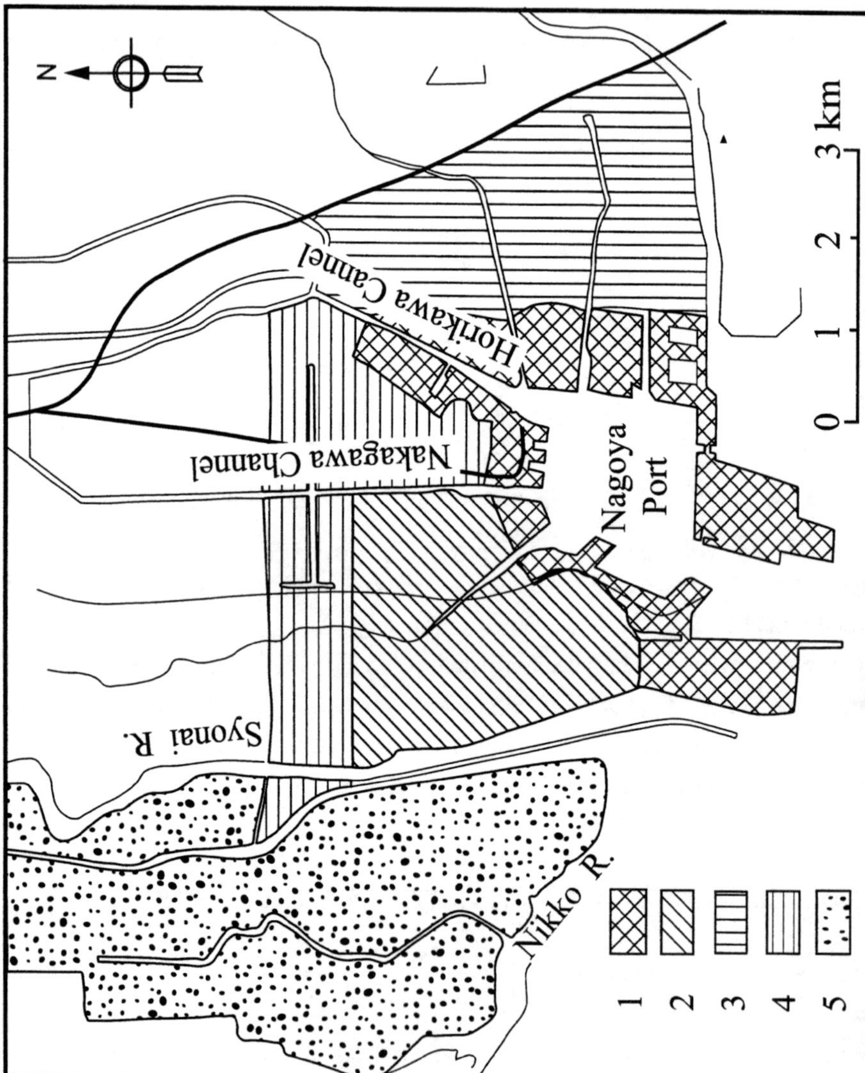

Figure 9.3 Construction zoning to aid flood control in Nagoya (Ministry of Construction, 1960; modified)
(1) 1st Area, (2) 2nd Area, (3) 3rd Area, (4) 4th Area, (5) 5th Area.

by rivers in the west and east, were to include refuge rooms at least 3.5 meters above ground level. Instructions for the third area, which consisted of reclaimed land for urban purposes, were similar to the second area. Buildings in the fourth area, which comprised older reclaimed land, and the fifth area, which consisted of reclaimed land for agriculture, were to have water-resistant structures at least 3.5 meters high.

CHAPTER 10

DETERMINING AREAS AT RISK OF SOIL LIQUEFACTION DURING AN EARTHQUAKE

Many past strong earthquakes have caused various types of damage to the ground, including faulting, formation of cracks, upheaval or subsidence, dislocation, landsliding and soil liquefaction accompanied by spouting water, sand and mud. In this section, the author discusses the determination of potential liquefaction sites through use of a geomorphological land classification map.

The effects of soil liquefaction can be dramatic and drastic, especially if it occurs extensively in a closely built-up area such as a city. It may even cause large-scale settlement or tilting of buildings. Various buildings in the city of Niigata were affected by liquefaction damage during the 1964 earthquake there, prompting structural engineers to try to establish countermeasures to protect structures from this type of ground damage. For this purpose, laboratory experiments such as soil vibration tests were carried out to determine what types of ground conditions were readily susceptible to soil liquefaction. In addition, sites in which soil liquefaction had already occurred were investigated, mainly by borings. Though such methods are effective, they are costly and time consuming.

10.1. Geomorphic Features of Sites Affected by Soil Liquefaction

Kazue Wakamatsu tried a different approach. She and her colleagues collected descriptions of past liquefactions from various old documents, and found that this kind of ground damage had occurred at various times and throughout the whole country. She then compared ground conditions at places where liquefaction had occurred with those where it had not and concluded that the potential for liquefaction at a site could be determined by a study of local ground conditions. Next, she applied a geomorphological land classification map to the study, as such a map shows both the micro-topography and the underlying ground conditions of a plain. In addition, ground conditions at points that had suffered liquefaction were investigated by borings (Wakamatsu, 1980, 1991).

In general, it was found that if areas are subjected to similar external forces the effects, such as deformation or displacement of the ground, are also similar. It also was found that liquefaction had recurred in the same regions during every large earthquake, and so the historical record of past liquefactions is valuable for predicting the potential for future occurrences of this type of ground damage.

The first attempt to determine potential soil liquefaction sites by using geomorphological land classification was done on the Shonai Plain in northern Honshu by Wakamatsu, Sumiko Kubo and the author in 1984.

In 1991, Wakamatsu compiled "Maps of Historic Liquefaction in Japan."

Photo 10.1 Soil liquefaction resulting from the Niigata Earthquake of 1964 caused this building to overturn.

Photo 10.2 The mounds in this paddy field were formed by soil liquefaction following the Niigata Earthquake of 1964.

Photo 10.3 A building sits on a large embankment along the Yodo River near Osaka (1990).

10.2. Flooding and Soil Liquefaction in the Shonai Plain

10.2.1. METHOD OF MAP PREPARATION

In preparing a geomorphological land classification map of the Shonai Plain, the area was first classified, using air photos, by major geomorphic elements such as mountains, terraces, valley plains, fans, natural levees, back-marshes, former river courses, deltas and sand dunes. This initial map, with a scale of 1:50,000, was then put into its next form by using the results of field surveys. Then, using the records of features caused by past flooding and soil liquefaction, and also air photos taken directly after the 1964 earthquake, a map was prepared showing where that quake had caused flooding and soil liquefaction. The group then used the map to study the relationships between the geomorphology of the alluvial plain and features of inundation and of soil liquefaction, and estimated the potential for future inundation and liquefaction in various areas.

10.2.2. BRIEF DESCRIPTION OF THE SHONAI PLAIN

The Shonai Plain, which faces the Japan Sea on the west, is triangular in shape (*Figure 10.1*). It is about 50 kilometers long to the northwest, about 5 kilometers wide in its northern part and 16 kilometers wide in the southern part. Although much smaller than the Kanto or Nobi Plain, it is extremely important because of its large rice production. The geomorphological land classification map of the region shows that the main Mogami River flows to the sea northwest across the central part of the plain, and the smaller Nikko, Gakko and Aka rivers flow seaward across the northern and southern parts of the plain.

In summary, the geomorphic features of the plain are as follows:

a) The alluvial fan is narrow along the main Mogami River, but wide along the Aka, Nikko and Gakko rivers.

b) The main Mogami River lacks a natural levee and a back-marsh.

c) Most of the plain is occupied by a flat and swampy delta, formed in a lagoon.

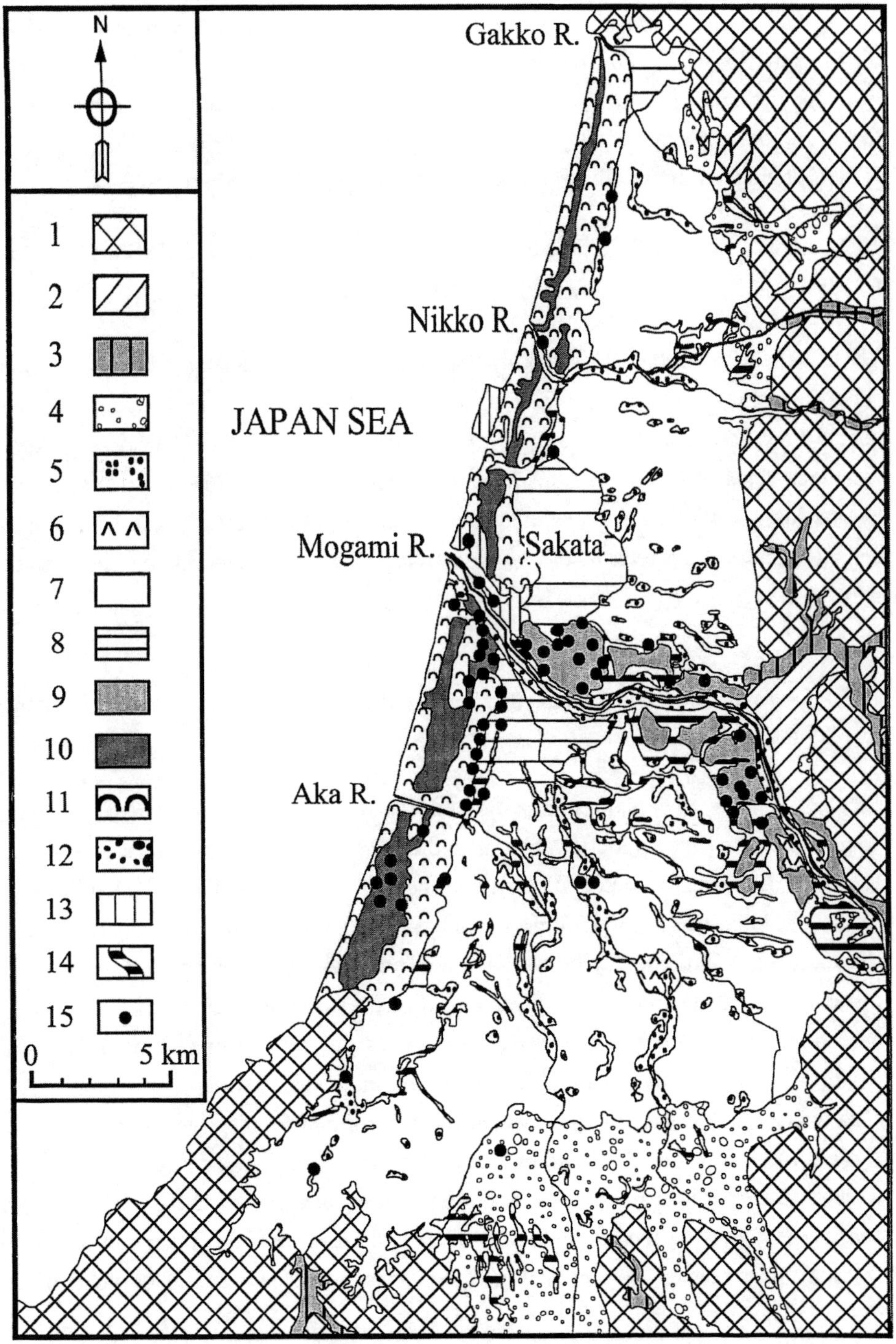

Figure 10.1 Geomorphological land classification and liquefaction sites in the Shonai Plain (Oya et al., 1982; modified) (1) Mountain or hill, (2) Terrace, (3) Valley plain, (4) Alluvial fan, (5) Natural levee, (6) Topography formed by bank collapse, (7) Upper delta, (8) Lower delta, (9) Lowland formed by meandering action of the Mogami River, (10) Lowland between sand dunes, (11) Sand dune, (12) Sand bar or riverbed, (13) Reclaimed land, (14) Former river course, (15) Site of liquefaction from the 1964 Niigata Earthquake.

d) The delta is divided into three parts: an upper delta, a lower delta and a lowland along the Mogami River formed by changes in its course.
e) To the west, the plain is bordered by a flat shoreline, with parallel belts of sand dunes that are among the longest in Japan.
f) Fields built on artificial fill are found near the mouth of the Mogami River.

10.2.3. RELATIONSHIP BETWEEN SOIL LIQUEFACTION CAUSED BY THE 1964 EARTHQUAKE AND GEOMORPHIC UNITS OF THE PLAIN

The 1964 Niigata Earthquake shook the whole Shonai Plain, causing many liquefaction sites. These, however, were restricted to the following units: a) a meandering former river course along the main river, b) a former river course beside the sand dunes and c) some reclaimed land and artificially filled fields near the estuary. The 1804 Kisagata and 1894 Shonai earthquakes also produced liquefaction sites on the plain. Although the locations and magnitudes of those earthquakes were different, the liquefaction occurred in the same units (*Figure 10.1*).

10.3. Hyogo Prefecture Nanbu Earthquake of 1995 (The Great Hanshin Earthquake)

On Jan. 17, 1995, an earthquake of magnitude 7.2 rocked Kobe and vicinity, and strong displacement took place along a fault line in the northern part of Awaji Island near Kobe. Heavy damage resulted, due to the epicenter being directly under the Kobe region. This earthquake killed about 6,400 people, gas and water pipe lines were destroyed, fires broke out and many buildings were burned, and telephone and telegraph communication was disrupted. Elevated parts of the highway and Shinkansen railway collapsed.

In addition, soil liquefaction caused embankments along the lower reaches of the Yodo River, which runs through Osaka, and along the Ina River to collapse in many places. This damage was especially great where the present and former river courses crossed (*Figure 10.2*). Along the Yodo River, lateral displacements from the liquefaction caused riverside walls to lean into the river. Ground water then gushed into protected areas, causing people living there to deduce that the embankment of the Yodo had failed. In addition, major lateral flow caused by the liquefaction took place in artificial fill at

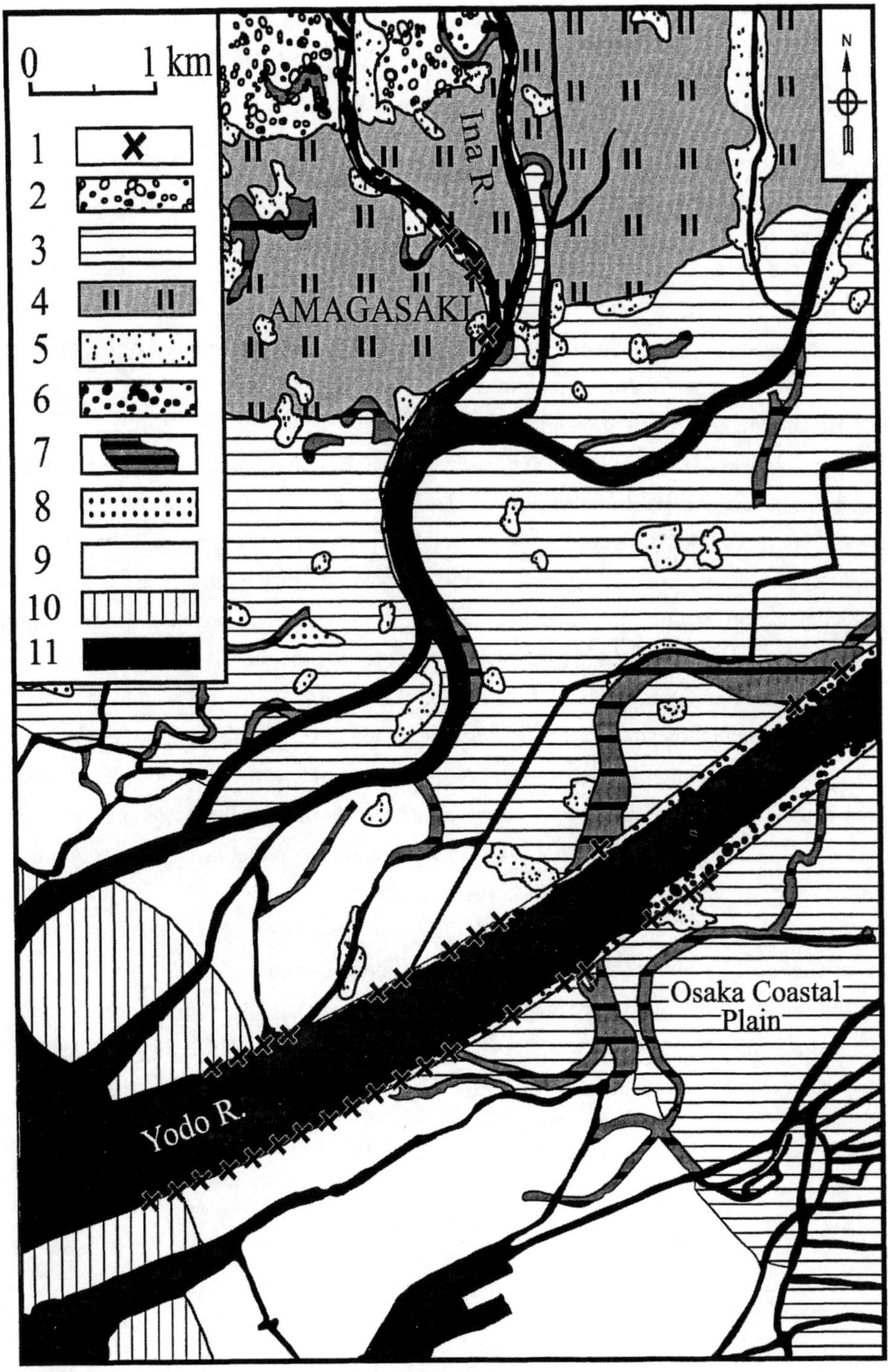

Figure 10.2 Geomorphological survey map of the Yodo River basin showing classification of flood-stricken areas following the 1995 earthquake (Oya and Kubo, 1993; modified)
(1) Destroyed embankment, (2) Fan, (3) Delta, (4) Back-marsh, (5) Natural levee, (6) Dry riverbed, (7) Former river course, (8) Sand spit, (9) Reclaimed land, (10) Artificially filled field, (11) River or sea.

the place known as Port Island, with maximum displacement of about 5 meters. This caused sea walls to fail, allowing much sand to be washed into the sea, which in turn caused substantial surface sinking farther inland.

10.4. Estimating the Potential for Soil Liquefaction from Earthquakes

One year after the Niigata Earthquake, the 1965 Nihonkai Chubu Quake, centered in the middle of the Japan Sea, rocked the Shonai Plain. It also caused soil liquefaction along the same former river course located beside the coastal sand dunes, in the northern part of the plain. This shows that locations of potential soil liquefaction caused by earthquakes can be determined by use of geomorphological maps.

Wakamatsu and the author also studied many other important plains in Japan, including the Kanto, Nobi and Fukui. Their studies clarified the following points:

a) The characteristics of liquefaction in an alluvial plain are influenced by its geomorphic features, which are peculiar to each plain. For example, in a relatively large plain that includes various geomorphic elements, such as the Nobi and Kanto plains, liquefaction sites vary according to whether the quake epicenter is inland or at sea, or whether in a lowland or mountainous district. On the other hand, in smaller plains such as the Shonai or Fukui, liquefaction sites tend to be more restricted to certain geomorphic elements.

b) In each alluvial plain, a close relationship exists between the various geomorphic elements and soil liquefaction. By clarifying that relationship, the possibility of future occurrences may be predicted for each element, and the predictions will become more accurate as features of the elements become clearer.

c) The potential for liquefaction in each element mentioned above apparently varies according to the geomorphic units of an alluvial plain, as noted in *Table 10.1.*

The above relationships were commonly observed, for example, in the Nihonkai Chubu Earthquake of 1983 (M=7.7), the Chibaken Toho Oki Earthquake of 1987 (M=6.7) and in the Great Hanshin Earthquake of 1995 (M=7.2).

Geomorphic elements	Possibility of liquefaction from seismic ground motion
Terrace	None
Valley plain	Low
Fan	Low in general, but exists to some degree in the underground spring zone along the lower edge
Natural levee	Exists to some degree in marginal parts
Back-marsh	Low in general, but exists to some degree
Delta	Low in general
Former river course	High
Sand spit	Low
Sand dune	Low in general, but high on the inland skirts of the sand dune zone
Lowland between sand dunes	Exists, especially high around underground springs
Reclaimed land	Low
Artificially filled field in shallow sea	High

Table 10.1 Relationships between geomorphic elements and possibility of liquefaction from seismic ground motion (Oya, 1995; modified)

CHAPTER 11

USE OF GEOMORPHOLOGICAL LAND CLASSIFICATION MAPS IN TECHNICAL ASSISTANCE TO DEVELOPING COUNTRIES

The usefulness of geomorphological maps for determining areas subject to flooding is recognized both in Japan and in developing countries, especially those in Southeast Asia that have much rain with resultant flooding. At the request of ECAFE, the author prepared such maps for Vientiane, Laos, and for Nongkhai, Thailand, and vicinity. Also, at the request of JICA, the author prepared a map for the Padang area of West Sumatra in Indonesia, and with assistance from the National Research Institute for Earth Science and from the Disaster Prevention Science and Technology Agency, he and others prepared a similar map for the Central Plain of Thailand.

Such maps are commonly used to study flooding and mitigation of damage from flooding. However, the author has also used them for other purposes, and will later discuss their use in selection of a bridge site in Bangladesh.

11.1. Use of a Geomorphological Land Classification Map in the Padang Project

The city of Padang lies just south of the equator. Formerly, the city was limited to the lower reaches of the Arau River, but a recent increase in population to about 500,000 has caused it to expand to the lower reaches of the Kuranji and Air Dingin rivers, resulting in the problems listed below. At the request of JICA and the Ministry of Public Works of the Government of Indonesia, the author has tried to resolve these problems by using a geomorphological survey map showing classification of flood-stricken areas.

1) The valley of the Arau River, which flows through the eastern part of the city, is too narrow for flood discharge and needs to be widened. If that were done, would its new channel cross section be reduced by future deposition of sand and gravel?

2) The Dutch constructed a flood way, and to further regulate flood discharge two weirs were constructed at the bifurcation between the Arau River and the flood way. The flood way is well sited, but it is too small to handle a large flood and should be enlarged. As the flood way crosses a fan where deposition is significant, this could rebury the flood way.

3) Wasteland lies on the right bank of the Laras River, which is a branch of the Kuranji River. If its ground conditions are good, it should be used for urban expansion; otherwise it could be used for a retarding reservoir. How should this land be used?
4) Padang is located beside the Indonesian Sea, and coastal erosion is significant—a great problem for the city. Conservation work will affect the coast at Padang. Will such work quicken or slow coastal erosion?

In response to these questions, the author prepared “A Geomorphological Survey Map of Padang and Surrounding Area,” which classified the areas prone to flooding. Before creating that map, he first made a topographic base map on a scale of 1:20,000. The area was first classified, using aerial photographs, by major features such as mountains, pyroclastic flows, fans, natural levees, back swamps, sand spits, abandoned channels and coral reefs. As it is best to use a photographic scale slightly larger than the projected map, the author used aerial photographs of scale 1:15,000. The initial map was put into final form by checking it with the results of field surveys.

The author learned there is a remarkably well-developed alluvial fan along the Kuranji River, and smaller ones along the lower reaches of the Air Dingin River and the middle reaches of the Arau. Pyroclastic flows are prominent along the Air Dingin River, and back swamps and deltas are found between the coastal sand spits and lower edges of the fan. An extensive swampy area is surrounded by the pyroclastic flows, sand spits and the natural levee of the Laras River. There is a dead coral reef at the mouth of the Kuranji and a living one farther offshore at the river mouth (*Figure 11.1*)

These geomorphological studies enabled the author to answer the questions posed above as follows:
1) Growth of the Kuranji River fan narrowed the Arau River near Barupiauan, which is upstream in Padang. There are several knick points in the Arau River canyon where stream flow is rapid, and there are marked differences in fluvial action above and below them. In the canyon, coarse gravel enters from both banks, and is also stirred up from the riverbed. In the upper part of the canyon, the maximum diameter of the gravel is 85 centimeters with a roundness of 0.6; downstream in the canyon it coarsens to 210 centimeters with a roundness of 0.5.

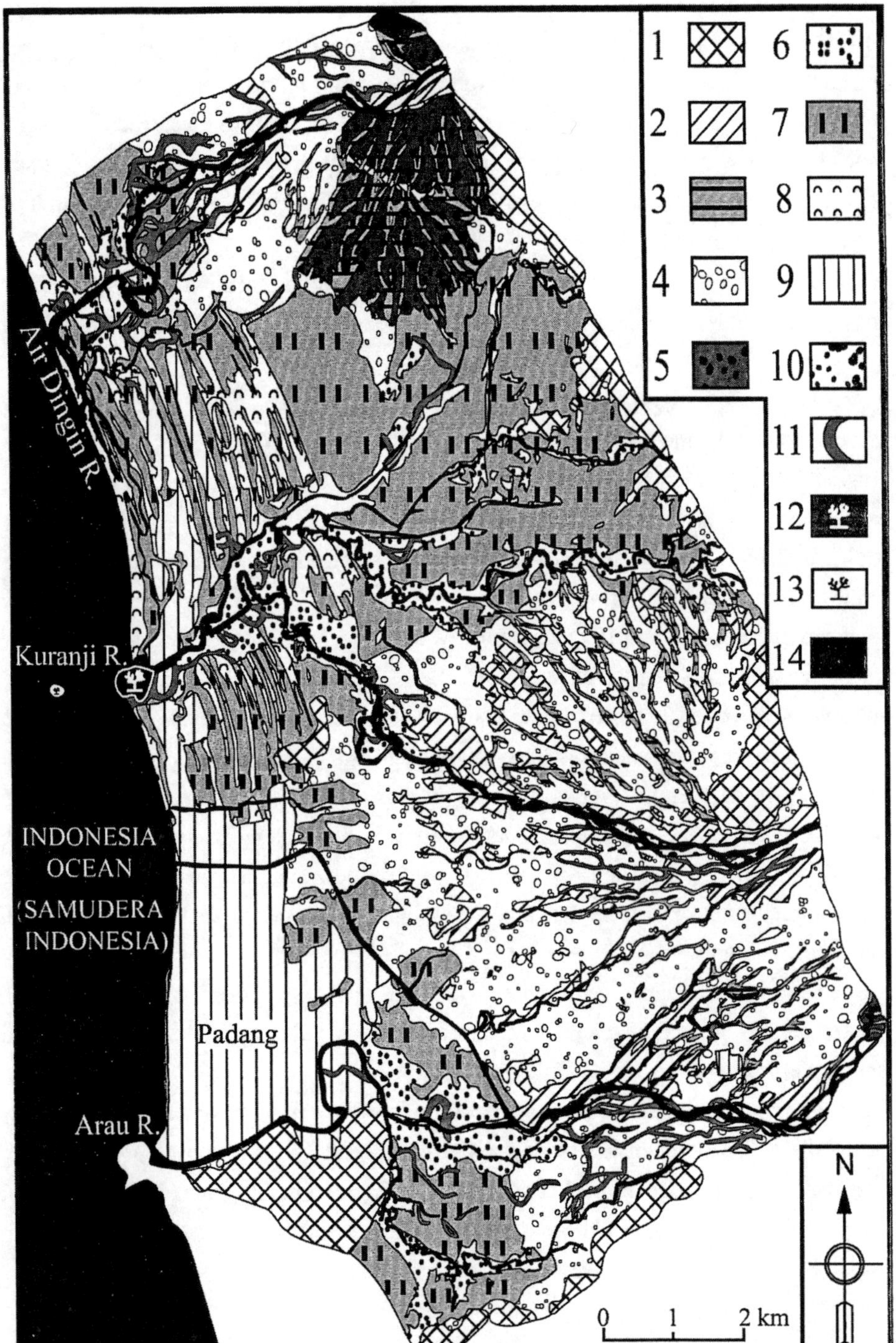

Figure 11.1 Geomorphological survey map of the Padang area in West Sumatra (Oya, 1983; modified) (1) Mountain or slope, (2) Terrace, (3) Valley plain, (4) Fan, (5) Pyroclastic flow terrace or volcanic fan, (6) Natural levee, (7) Back swamp, (8) Sandbank or sand dune, (9) Artificially filled field, (10) Dry riverbed, (11) Former river course, (12) Dead coral reef, (13) Living coral reef, (14) River or sea.

Gravel in the upper reaches of the canyon consists of andesite and granite, in a ratio of 10:1. However, granite is lacking farther downstream, and there also is none in the upper reaches of the Kuranji River. Furthermore, in its upper reaches the stream is braided. During floods, water from the river overflows adjacent areas. These facts indicate strong deposition in the upper reaches of the canyon but that much of the stream load is dropped at the knick point, with only fine gravel and sand carried beyond it, and so the stream load is much smaller there.

Altogether, the studies show that Padang has a good natural location. However, if river conservancy action is taken, engineers must pay attention to maintaining the narrows as at present.

2) Much of the Kuranji fan is slightly hilly, with many villages. It is dissected, especially in its upper part, by the main course of the Kuranji and by its tributaries. Material from the erosion of the fan by those streams has built a new fan between the original one and the sand spits and back swamps.

The author proposes that the Padang Plain developed as follows. Pyroclastic flows from large volcanic eruptions formed its base. Later, the pyroclastic material was covered by sand and gravel some 5 to 10 meters thick to form the surface of the fan. The age of the fan is uncertain, but it is probably of the Würm Ice Age, about 18,000 years ago, when the sea level was about 100 meters lower than now due to glacial eustasy. When the climate later warmed, between about 6,000 to 4,000 years ago, the sea rose and reached 2 to 6 meters above its present level, and the lower part of the fan was covered with marine deposits. Later, as the sea level fell, sand spits were built along its seaward side. Next, crustal movement forced the Kuranji River into a new channel, and its muddy water killed some of the coral reefs. Further, the fan and its pyroclastic material were eroded, and a new fan built.

The above facts indicate that fan deposition has ended and been replaced by erosion. Therefore, it is unlikely that an enlarged flood way will subsequently be filled with sand and gravel from the fan.

3) In addition to studying the topography of the wasteland, the author has also tested it by bore holes. The area consists of peaty marsh. As peat generally develops only in the Frigid Zone, and under tropical conditions weathers and disintegrates into mud, a past change of climate is indicated. As the peaty marsh is soft and unsuited for buildings, it would be economical to use the area for a retarding reservoir.

4) A coastal embankment surrounds the area, and more than 30 jetties have been constructed along it since 1969. Small deposits are seen south of each jetty. Local inhabitants and fishermen state that the coast has retreated about 25 meters since 1969.

Coconut palms are abundant in the area, but some have died due to salt damage and coastal erosion. The present sandy beach is widest at the mouth of the Kuranji, the biggest of the three rivers. There is no coastal erosion between the Kuranji River and the Air Dingin, as the sea is still retreating in that stretch.

The coastal deposits consist mostly of sand with minor gravel. The gravel is more rounded than normal beach gravel and resembles river gravel. It is composed of andesite. These features indicate that the gravel came down the Kuranji River.

The Padang coast shows many well-developed sand spits or sandbanks, which form about 20 lines along the coast. The width of this band of spits is 1.5 kilometers between the Kuranji and Air Dingin rivers, 1.5 to 1.2 kilometers between the Kuranji and the Lolong River, and about 0.8 kilometers from that river to the Arau. This indicates the coastal area has widened north from the Kuranji River and narrowed to the south.

The author considers that this difference, along with the coastal erosion south of the river mouth that contrasts to the deposition north of it, is caused by the dead coral reef at the mouth of the Kuranji River. That reef forces the river to swing slightly to the north at that point, dispersing in that direction the muddy river water entering the sea, where it lays down its sand and silt. This phenomenon can be observed both in aerial photographs and directly from the color of the seawater; north of the Kuranji the seawater is a light brown but to the south, where living coral reef can be found, it is light blue.

11.2. Map Use in Selecting a Site for a Bridge over the Brahmaputra-Jamuna River

After Bangladesh seceded from the Islamic Republic of Pakistan and became a sovereign state on March 29, 1970, many problems hindered its progress. It had a war-torn economy and the very low annual per capita income of U.S. $70. About 72.5 percent of its budget came from assistance from foreign countries such as the United States and India. It was evident that its success or failure would depend on such aid.

Many factors hindered the economic development of Bangladesh. First, its population continued to grow whereas rice production did not. Second, the education level remained low, partly due to the increasing population. Further, road and rail networking was poor. While there were many large rivers suitable for inland navigation, such as the Ganges River from the west, the Brahmaputra-Jamuna from the north and the Meghna from the east, these in turn obstructed road and rail networks. As there were few bridges, ferries were common, and crossing a river could take much time due to congestion. At that time, it could take about 20 hours to travel the 200-kilometer route between the capital, Dhaka, and the main industrial city, Chittagong, because of congestion and confusion at the ferry in the 1970s.

Railways were constructed during the British regime before World War II, but the large rise in water during flooding—as much as 11 meters—caused much difficulty. Also, during the dry season the Brahmaputra-Jamuna River is only about a fifth as wide as during the rainy season. As a result, during the dry season it was necessary to extend the railway on the sandbank in order to reach the harbor. Further, due to dry season lowering of the river level with consequent deposition of sand, the important rail line between Dhaka and Bogra via Mymensingh had to extend its track southward about 20 kilometers. Trains had to travel slowly over this stretch because of the soft ground.

After Bangladesh became independent, the government sought to unite the territories separated by the Brahmaputra-Jamuna River and asked Japan for assistance in bridging it.

First, a Japanese bridge expert roughly measured the width of the river on a map, and found it to be about the same as the distance across the strait of Japan's Seto Inland Sea, which was being bridged at the time. He therefore concluded that the river could be bridged. As a result, the government of Japan organized the Jamuna River Bridge Construction Project Team, led by Shizuo Inose and Seiichi Sato. The team found that bank erosion and shifting of river course were significant, and that because of the river's size its course could not be stabilized, as was done in Japan. Thus, it was necessary to find a stable location along the river for the bridge.

During the previous 20 years, the author had prepared geomorphological land classification maps for many plains in Japan and Southeast Asia, on which former river courses could be traced. Therefore, he was requested to prepare a similar map for this area in order to find a natural stable point along the river for the bridge site.

11.2.1. METHODOLOGY

At the time the author was asked to prepare the map, there were already maps on various scales, such as 1:25,000, 1:50,000 and 1:250,000, but they were not very precise. Also, stream bank erosion and shifting of river courses were fast and widespread. Therefore, the author prepared fresh geomorphological maps, using a scale of 1:1,000,000 for the Brahmaputra-Jamuna-Ganges River Plain map and a scale of 1:50,000 for that of Brahmaputra-Jamuna River (Sirajganj area). As a base, he used photo mosaics created from aerial photographs taken by the Earth Resources Technology Satellite (ERTS) during the dry season of 1972.

Infrared photos were used because they are best suited for geomorphic photo interpretation. Those photos show water surfaces such as lakes, rivers and sea in black, marshy areas such as back-marshes and tidal flats in a lighter shade of black, and dry areas such as terraces and natural levees in white. Using the ERTS photos, along with field observations by helicopter, boat and jeep, the region was classified into geomorphic elements such as: steep mountain slopes, gentle piedmont slopes, hills, upper and lower terraces, valley plains, gently sloping sandy alluvial fans, natural levees, former (higher) deltas, back swamps or deltas, deltas influenced by tides and former river courses (*Figure 11.2*).

The team took its own set of aerial photographs in 1974.

11.3. Geomorphological Features of the Brahmaputra-Jamuna-Ganges River Plain

Understanding the geomorphological features of an area is crucial in determining a stable point to construct a bridge; these features reveal what has happened in the past and will likely happen in the future. It is especially important for bridging a river such as the Brahmaputra-Jamuna-Ganges, which has a history of frequently changing its course.

The Brahmaputra-Jamuna-Ganges River Plain is surrounded by the Shillong Plateau to the north, the Chittagong Hills in the east, Rajimahal Hills in the west and the Bay of Bengal to the south. The Chittagong Hills are fold mountains that strike north, and the many plains that occupy the valleys between the hills parallel them. This has allowed construction of large dams in the valleys between the hills.

There are Pleistocene terraces between the Jamuna and Meghna rivers and between the Jamuna and the Ganges. The former has been named Maduhupur Jungle Terrace (Morgan and McIntire, 1959), and the latter Western Barind Terrace. The first, also known as Eastern Barind Terrace, is triangular in shape and lies about 10 to 17 meters above sea level. It can be divided by height into an upper and a lower terrace. The city of Dhaka lies on the lower part of the terrace. The western boundary between the terrace and the adjoining alluvial plain is a 10-meter-high cliff. The boundary is less clear on the east,

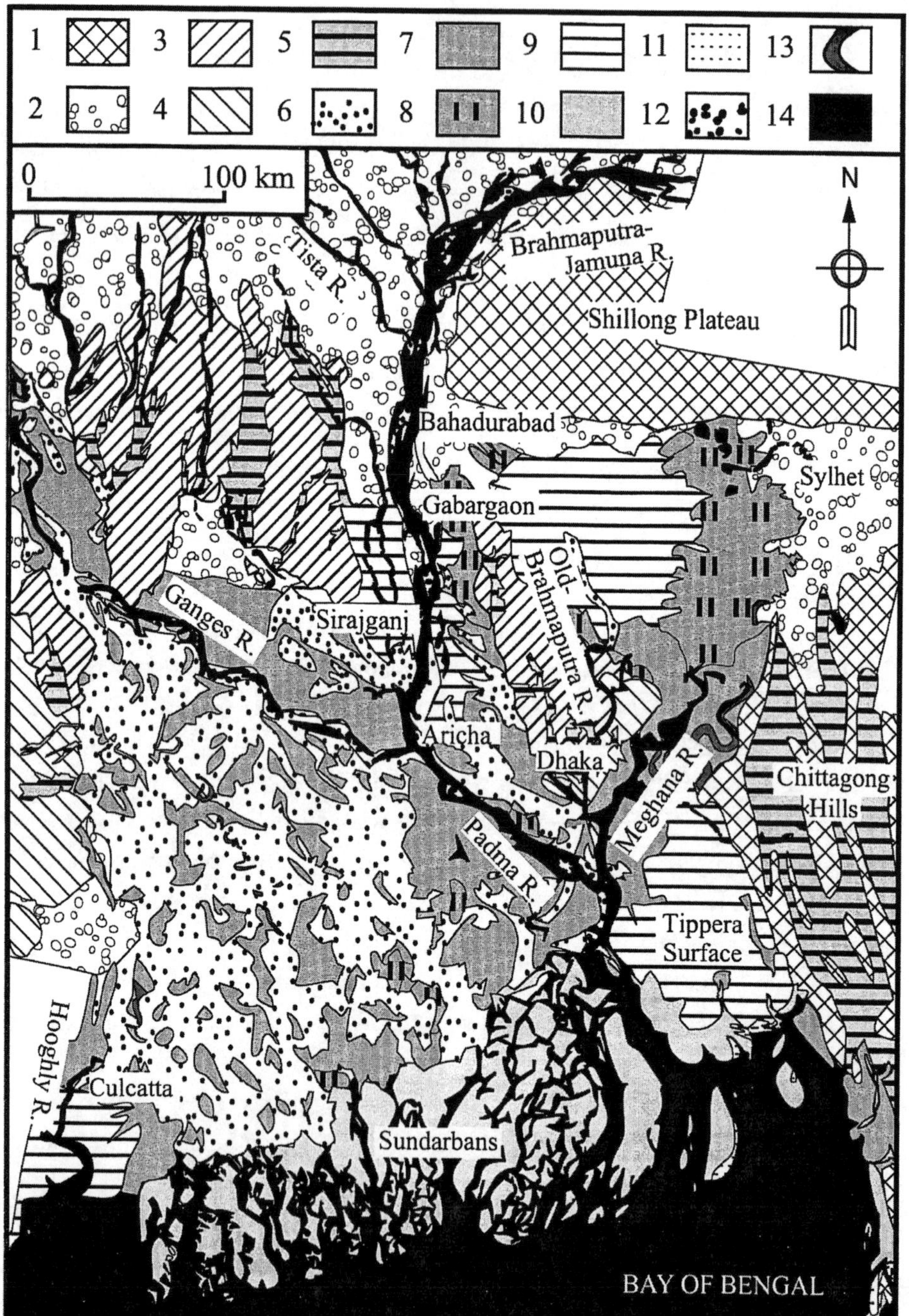

Figure 11.2 Geomorphological map of the Brahmaputra-Jamuna R. and Ganjes R. Plain (Oya, 1976; simplified) (1) Mountain or hill, (2) Sandy alluvial fan with gentle slope, (3) Upper terrace, (4) Lower terrace, (5) Valley plain, (6) Natural levee, (7) Back-swamp or delta, (8) Marsh, (9) Former or upper delta, (10) Delta influenced by tide, (11) Tidal flat, (12) Dry riverbed, (13) Former river course, (14) River or sea.

Photo 11.1 The Brahmaputra-Jamuna River in Bangladesh (1974)

Photo 11.2 Sylhet basin in Bangladesh (1974)

where the terrace grades into an alluvial plain. The topography and the results of the studies by James Morgan and William McIntire caused the author to consider the terrace to be a tilt block.

Both the Western and Eastern terraces are fan shaped, but the western one is larger and several rivers, including the Atrai and Karatoya, dissect it. Whereas the soil of the Maduhupur Jungle Terrace consists of red clay, that of the Western Barind Terrace is sandy. Because of the above features, the author considers that the former is an upheaved delta, but the Western Barind Terrace is a gently sloping, upheaved fan.

There are also terraces in the northern part of the Chittagong Hills and in the eastern part of the Rajimahal Hills, on the western side of the Hooghly River, which flows through Calcutta. Apart from the mountains, hills and terraces mentioned above, Bangladesh rests on an alluvial plain.

A small alluvial fan, with a radius of only about 20 kilometers, extends the Shillong Plateau. It consists of sand and gravel from its apex to near its center, but from there to its lower edge only of sand. As the border between India and Bangladesh lies in the center of the fan, there is no gravel in the Bangladesh part, adding a further obstacle to bridge construction.

Alluvial fans cover the northern part of the Western Barind Terrace. These were formed by the Tista River, which rises in the Himalaya Mountains. Most of the terrace consists of sand, and it has a gentle slope of about 1/10,000, partly due to its paucity of gravel—a result of the tropical climate and of its great distance from the Himalayas. The fans thus differ from ones in Japan, which consist of both sand and gravel and have steep slopes.

The Brahmaputra-Jamuna River's flow pattern differs between its upper and lower reaches. In the former it resembles a net, a pattern typical of an alluvial fan, but in its lower reaches it displays a serpentine pattern, typical of a natural levee region. These features indicate the river has advanced about 50 kilometers beyond the protected area surrounding the city of Bogra.

The alluvial plain bordering the southern part of the Ganges River was formed mostly by deposits from that river.

A former delta is located in the lower reaches of the Old Brahmaputra River near Sirajganj. The Old Brahmaputra, Tista and other rivers built it prior to 1830.

The city of Comilla lies on an upper delta, which Morgan and McIntire (1959) named "Tippera Surface." Located between Meghna River and the Chittagong Hills, this delta is considered to be of Holocene Age.

Facing the Bay of Bengal is a delta area that is strongly affected by tides. Residents call it "Sundarbans," which means beautiful forests of Sundri Trees (Heritiera Fomes). The area has dense forests of mangrove, nipa palm, etc., and it contains many animals, including tigers, deer, crocodiles, eagles and macaws.

A large marshy area, the Sylhet Depression, occupies a low area to the southwest of the city of Sylhet. The average elevation of the depression is 5 meters above sea level, with its lowest part only 3 meters, even though it is 180 kilometers from the coast. Drainage in the area is poor, largely due to its low elevation. As well, the depression's southern part is blocked by the Maduhupur Jungle Terrace and Tippera Surface, the Meghna River is partly obstructed by the natural levee of the lower section of the Jamuna River (Padma River), and the Meghna River's gradient is low; it is only about 5/100,000 at the Meghna Bridge (*Figure 11.2*).

During the rainy season, water levels in the depression can reach 6 meters. These conditions are ideal for floating rice, and it is the area's main crop. The rice stems are among the longest in the world.

11.4. Geomorphic Development of the Brahmaputra-Jamuna-Ganges River Plain

The sea level was lowered about 100 meters during the Würm Ice Age. Further, low temperatures at that time caused severe mechanical disintegration of rock in the mountains, producing a large quantity of sand and gravel that formed big fans at the foot of the Himalayas and the Shillong Plateau.

In 1976, geologists of the Japanese Jamuna River Bridge Construction Project Team put down four bore holes for geological purposes along the Jamuna River, at Bahadurabad, Gabargaon, Sirajganj and Aricha. They discovered a gravel layer about 10 meters below the surface at Sirajganj, and at a depth of 70 meters at Aricha. In both places it was covered by silt and sand. A carbon date of 28,000 years on fossil wood from below the gravel at Sirajganj indicates that the gravel was deposited during the Würm Ice Age. In contrast, in Japan the gravel is usually about 40 to 70 meters below the surface, and is overlain by blue marine clays, which are absent here. The difference may be due to the greater distance that the Bangladesh bore holes are from the sea, which caused a predominance there of fluvial, rather than marine, deposits during the Holocene.

The area is in the Himalayan Orogenic Movement Region, and suffered earthquakes in 1762, 1775, 1812 and 1912, with each large quake causing changes in ground level. The Maduhupur Jungle and the Western Barind Terraces were formed during the Pleistocene and the Tippera Surface during the Holocene, and they, along with the Ganges Plain, contain many elevated areas. The longitudinal direction of the intervening sunken areas is parallel to the strike of the Chittagong Hills. However, the Ganges Plain contains many fault lines that strike northeast, or almost perpendicular to the strike of the hills.

The alluvial plains are located between terraces, or between a terrace and a hill or mountain, and they and the delta area have been depressed. There are also many small rivers in the delta west of the Haringhata River, which dissects that region. Buried fossil wood in the Sylhet Depression indicates that changes in ground level continue.

11.5. Shifting of the Ganges and Brahmaputra-Jamuna Rivers

11.5.1. SHIFTING OF THE GANGES RIVER COURSE

There has been marked shifting of course of both the Ganges and Brahmaputra rivers in the plain, resembling the shifting of the Yellow River in China. It has been caused largely by strong crustal movements, the large annual fluctuation of flood discharge, the considerable volume of the river deposits and because there has been no attempt to stabilize the river courses.

There are two theories regarding the shifting of the course of the Ganges River. One assumes that the Hooghly River now occupies the former course of the Ganges River, and that it gradually shifted from west to east. The other considers that both rivers existed from ancient times but their size changed. No matter which theory is correct, it is evident that crustal movements have influenced the shifting. In general, the size of the Hooghly River has gradually decreased, due to a lessening of water flow from the Ganges River caused by rising ground.

11.5.2. SHIFTING OF THE BRAHMAPUTRA-JAMUNA RIVER COURSE

A map prepared by Major James Rennell, who studied the Bengal region from 1764 to 1773, portrays the Brahmaputra-Jamuna River running through Mymensingh and along the Old Brahmaputra course. He also shows a small stream following the present course of the Jamuna River, which was called Jhinai River in its upper reaches and Jamuna in its lower. In 1916, Major F.C. Hirst suggested that the shift from that course to its present one had taken place gradually between 1720 and 1830, or during the 50 years following publication of Rennell's map. However, Buchanan Hamilton reported in 1810 that the Jamuna River was on the verge of shifting (Morgan and McIntire, 1959).

Opinions differ as to what caused the shifting of the Brahmaputra-Jamuna River.

J. Fergusson considered that it was caused by crustal movement, i.e. the upheaval of the Maduhupur Jungle Terrace and sinking of the Jamuna Valley plain. On the other hand, La Touche, H.H. Hayden, E.H. Pascoe and others proposed that the Dihang River, a tributary to the Brahmaputra in its upper reaches, beheaded the Tsang Po River in the Tibetan Plateau. This suddenly increased the discharge in the Brahmaputra-Jamuna, causing part of its flow to pour into the Jhinai River, which therefore became larger (Morgan and McIntire, 1959).

The author supports Fergusson's theory. In 1976, the author confirmed that the northern part of the Maduhupur Jungle Terrace could be found along the north bank of the Old Brahmaputra River, showing that upheaval of the Maduhupur Jungle Terrace blocked the Brahmaputra-Jamuna River. That upheaval, along with the shifting of the Tista River and the many sandbanks formed from the big load it carried, caused the courses of the Brahmaputra and Jhinai rivers to be devastated by floods, and the Jhinai River expanded to become the main branch of the Brahmaputra-Jamuna River.

Later, the Brahmaputra River shortened its course to the Bay of Bengal. Nowadays, the maximum flood flow of the Old Brahmaputra River is only about 1/40 of that of the present Brahmaputra (Jamuna) River, and in the dry season its discharge is meager.

11.6. Selection of the Bridge Site

The river engineers selected Bahadurabad, Gabargaon, Sirajganj and Aricha as possible bridge sites. The author was requested to select the most suitable of those sites for a crossing, based on his research. Using the geomorphological map, the author found that the Bahadurabad and Gabargaon sites lay on the alluvial fan, Sirajganj on a natural levee and Aricha in the delta region (*Figures 11.3, 11.4*). Generally speaking, rivers on alluvial fans frequently shift courses, whereas those in natural levees are more stable. The delta region where Aricha is located is strongly affected by flow of the Ganges.

There is a narrow ridge at Sirajganj that was formed by the old alluvial plain. Small rivers formed this plain about 180 years ago when the Brahmaputra-Jamuna River shifted into this area, and many villages more than 200 years old lie on it. These facts indicate that the ridge area has been stable for at least 180 years.

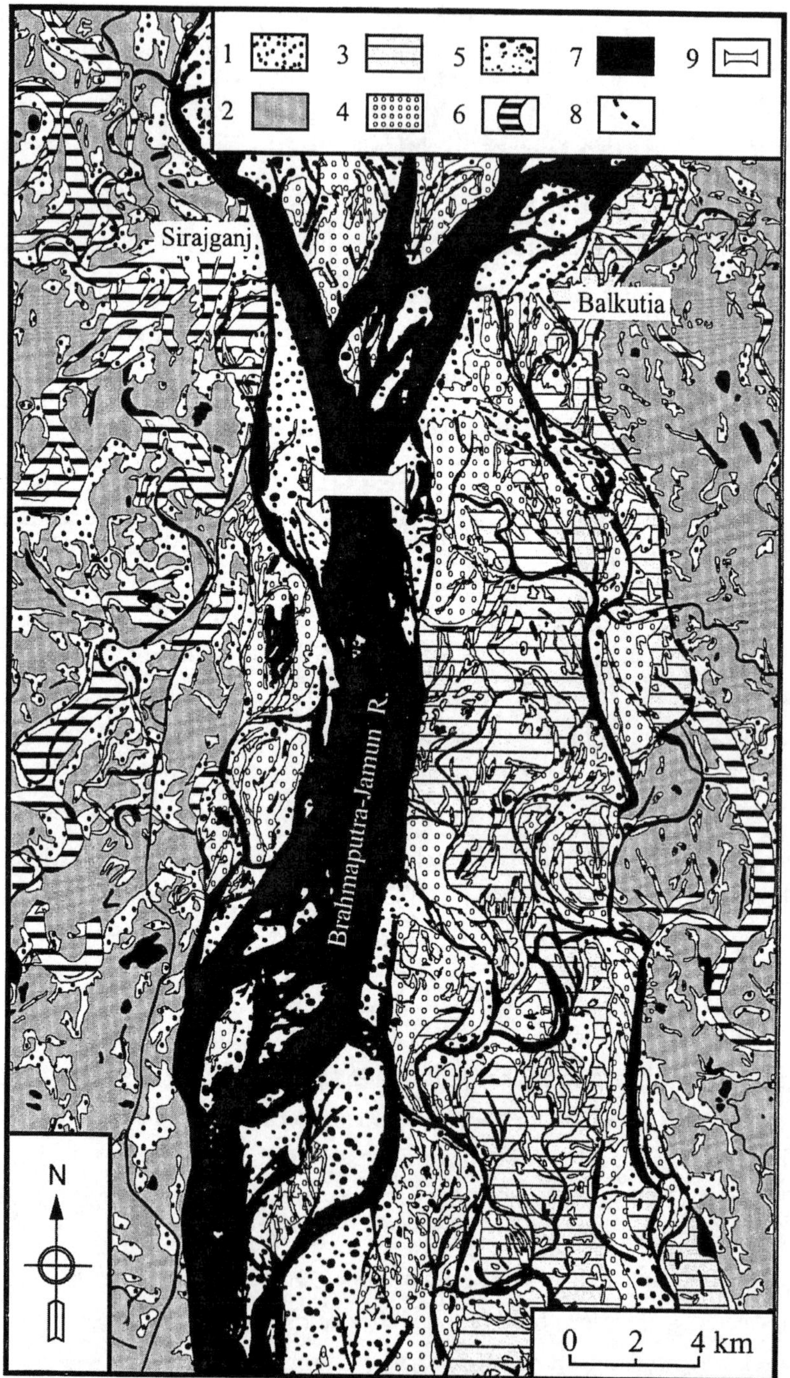

Figure 11.3 Geomorphological land classification map of the Brahmaputra-Jamuna River basin, Sirajganj area (Oya, 1976; modified)
(1) Natural levee, (2) Back-marsh, (3) Delta, (4) Sandbank covered by vegetation, (5) Dry riverbed, (6) Former river course, (7) River, (8) Cliff, (9) Bridge Site.

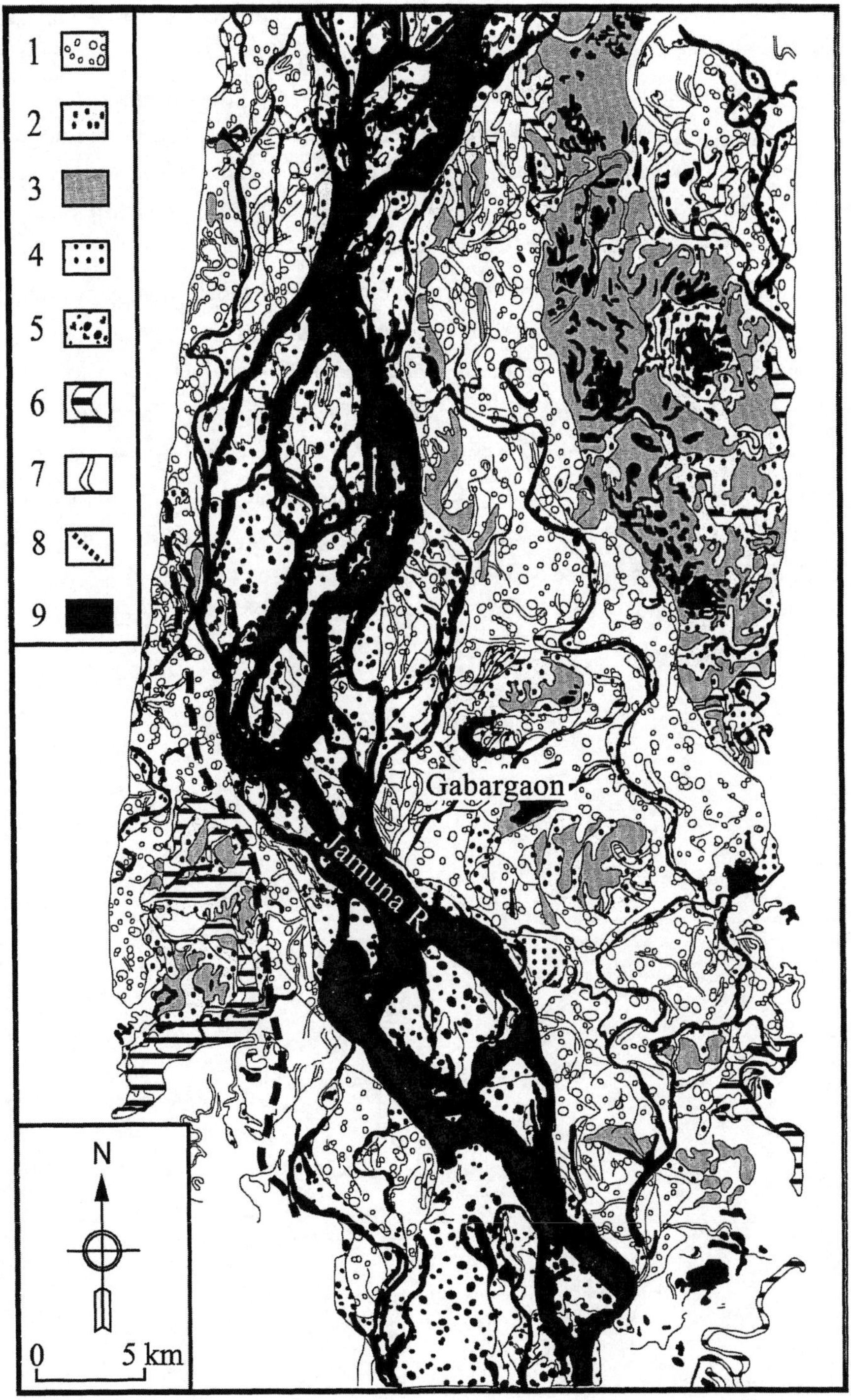

Figure 11.4 Geomorphological land classification map of the Brahmaputra-Jamuna River basin, Gabargaon area (Oya, 1976; modified)
(1) Fan, (2) Natural levee, (3) Back-marsh, (4) Point bar, (5) Dry riverbed, (6) Abandoned river channel, (7) Meander scroll, (8) Cliff, (9) River.

A cliff lies between the old and new alluvial plains along the Brahmaputra-Jamuna River. It is highest, at about 4 meters, opposite Sirajganj but lowers to the north and south and eventually disappears. Downstream from Sirajganj, the direction of the former river courses is from northeast to southwest on the left bank and from northwest to southeast on the right. These features show that the line connecting Sirajganj with the opposite side of the river is located on the geoanticlinal belt, and that the narrow ridge at Sirajganj was formed by rising ground.

Considering the process that formed it, the author believes that the ridge will continue to be stable. Though many faults, some active, cross this plain, there is none near Sirajganj. Therefore, from the geomorphological point of view, the author regarded the Sirajganj site as the best among the four locations for construction of a bridge.

Other details regarding geology, traffic, cost of construction and so on were taken into account by the Jamuna River Bridge Construction Project Team. The results of their findings tallied with the author's own conclusions from the geomorphological research, and it was decided to build the bridge at Sirajganj.

In 1998, 20 years after the author's research, the bridge was completed at the recommended site (*Figure 11.3*).

APPENDIX

Area	Year Published	Map Title	Author	Publishing Office
Kiso River (Nobi Plain)	1956	Reconnaissance Topographical Survey Map of the Lower Part of the Kiso River Basin	M. OYA	Resources Council
Chikugo River	1957	Topographical Survey Map of the Chikugo River Basin Showing Classification of Flood-Stricken Areas	M. OYA	Resources Bureau, Science and Technology Agency
Isahaya	1959	Topographical Survey Map of the City of Isahaya and Vicinity Showing Classification of Flood-Stricken Areas	M. OYA	Resources Bureau, Science and Technology Agency
Ishikari River	1959	Topographical Survey Map of the Ishikari River Showing Classification of Flood-Stricken Areas	M. OYA	Resources Bureau, Science and Technology Agency
Naka River	1961	Geomorphological Classification Map of a Section of the Naka River Basin	F. TADA M. OYA H. ONO Y. KAWASHIMA	Resources Bureau, Science and Technology Agency

Appended Table 1 Geomorphological Survey Maps Showing Classification of Flood-Stricken Areas

Area	Year Published	Map Title	Author	Publishing Office
Sanriku Coast	1962	Topographical Survey Map of Kesenuma and Vicinity Showing Classification of Tsunami-Stricken Areas Map Showing Classification of Micro-topography of Shizugawa and Vicinity Map Showing Classification of Topography of the Shiogama Harbor and Vicinity Showing Tsunami-Types of the Region	T. NAKANO M. IKEDA K. YAMAGUCHI M. OYA Y. MATSUYAMA K. KIKUCHI	Coast Section, River Bureau, Ministry of Construction, Geographical Survey Institute, Ministry of Construction
Ariake Coast	1963	Topographical Survey Map of the Ariake Kai Basin (Kyushu) Showing Classification of Flood-Stricken Areas	M. TAKASAKI M. OYA M. NAGASE K. KIKUCHI	Geographical Survey Institute, Ministry of Construction
Yoshino River	1963	Geomorphological Survey Map of the Yoshino River Basin Showing Classification of Flood-Stricken Areas	M. OYA	Resources Bureau, Science and Technology Agency
Kano River	1966	Geomorphological Survey Map of the Kano River Basin Showing Classification of Flood-Stricken Areas	M. OYA	Resources Bureau, Science and Technology Agency
Kuzuryu River	1968	Geomorphological Survey Map of the Kuzuryu River Basin Indicating Areas Subject to Flooding	M. OYA	Resources Bureau, Science and Technology Agency

Area	Year Published	Map Title	Author	Publishing Office
Nakatsu River	1971	Geomorphological Survey Map of the Nakatsu River Basin Situated in the Upper Reaches of the Kiso River	M. OYA	Nagoya Regional Forest office
Neya River (Osaka Plain)	1972	Geomorphological Land Classification Map of the Neya River Basin (Osaka and the Surrounding Area) Indicating Areas Subject to Flooding	M. OYA	National Research Center for Disaster Prevention Science
Tsugaru Plain	1977	Geomorphological Survey Map of the Tsugaru Plain Indicating Areas Subject to Flooding	F. TADA M. OYA M. UMITSU et al.	Ministry of Education
Shizuoka Plain	1977	Geomorphological Survey Map of the Shizuoka Plain Showing Classification of Flood-Stricken Areas	M. OYA	Civil Engineering Research Institute, Ministry of Construction
Yahagi River	1977	Geomorphological Survey Map of the Yahagi River Plain Indicating Areas Subject to Flooding	M. OYA S. SUGIURA	River Work Office at Toyohashi, Ministry of Construction

Area	Year Published	Map Title	Author	Publishing Office
Agano River	1984	Geomorphological Survey Map of the Agano River Basin Showing Classification of Flood-Stricken Area	M. OYA Y. KATO	River Work Office at Aganogawa, Ministry of Construction
Tokoro River	1985	Geomorphological Survey Map of the Tokoro River Basin Showing Classification of Flood-Stricken Area	M. OYA M. UMITSU S. HARUYAMA Y. HIRAI	Division of Development and Construction at Abashiri, Bureau of Hokkaido Development
Kise River	1985	Geomorphological Land Classification Map of the Kise River Basin	M. OYA K. KOTODA K. WAKAMATSU Y. TAKAGI et al.	Work office of Numazu, Ministry of Construction
Kasumigaura Lake, Kitaura Lake	1986	Geomorphological Land Classification Map of Kasumigaura and Kitaura Lakes	M. OYA Y. KATO S. HARUYAMA Y. HIRAI et al.	Work office of Kasumigaura, Ministry of Construction
Katsushika Ward	1985	Geomorphological Land Classification Map of Katsushika Ward (Tokyo)	M. OYA S. HARUYAMA	Board of Education of Katsushika Ward

Area	Year Published	Map Title	Author	Publishing Office
Sagami Bay	1991	Geomorphological Land Classification Map of the Northern Coast of Sagami Bay	M. OYA A. MATSUBARA S. KUBO K. KODERA	Keihin Work office, Ministry of Construction
Yodo River	1993	Geomorphological Survey Map of the Yodo River Basin Showing Classification of Flood-Stricken Areas	M. OYA S. KUBO	Yodo River Work Office, Ministry of Construction
Naruto Straits	1994	Geomorphological Map of the Coast and Sea Bottom of Tokushima Prefecture (River Mouth of the Yoshino and Straits of Naruto)	M. OYA S. HARUYAMA et al.	Tokushima Prefecture
Yoshino River	1995	Geomorphological Survey Map of the Yoshino River Basin, Shikoku, Showing Classification of Flood-Stricken Areas	M. OYA S. HARUYAMA Y. HIRAI A. MATSUDA	Tokushima Work Office, Ministry of Construction
Ara River	1996	Geomorphological Survey Map of the Ara River	M. OYA H. TAKAYAMA S. KUBO	Arakawa Joryu Construction Office, Ministry of Construction
Naka River	1997	Geomorphological Survey Map of Naka River Basin, Shikoku, Showing Classification of Flood-Stricken Areas	M. OYA S. HARUYAMA Y. HIRAI	Tokushima Work Office, Ministry of Construction

Area	Year Published	Map Title	Author	Publishing Office
Mune River (Thailand)	1961	Topographical Survey Map of the Mune River Showing Classification of Flood-Stricken Area	M. OYA	Mekong Reconnaissance Team, Government of Japan
Gum River (Thailand)	1961	Topographical Survey Map of the Gum River Showing Classification of Flood-Stricken Area	M. OYA	Mekong Reconnaissance Team, Government of Japan
Mekong River Vientiane Plain (Laos and Thailand)	1967	Geomorphological Survey Map of the Mekong River Basin (Vientiane, Nongkhai and the Surrounding Area situated immediately down stream from Pa Mong) Indicating Areas Subject to Flooding	M. OYA	Mekong Committee
Brahmaputra -Jamuna River (Bangladesh)	1976	Geomorphological Land Classification Map of the Brahmaputra-Jamuna River Basin, Sirajganj Area Geomorphological Land Classification Map of the Brahmaputra-Jamuna River Basin, Gabargaon Area	M. OYA	JICA
Ganges Plain (Bangladesh)	1976	Geomorphological Map of the Brahmaputra-Jamuna River and Ganges River Plain	M. OYA	JICA

Area	Year Published	Map Title	Author	Publishing Office
Toyo River	1978	Geomorphological Survey Map of the Toyo River Basin Showing Classification of Flood-Stricken Areas for Flood Control	M. OYA F. OMORI	Work Office at Toyohashi, Ministry of Construction
Shonai River	1979	Geomorphological Survey Map of the Shonai River Basin Showing Classification of Flood-Stricken Areas for Flood Control	M. OYA S. SUGIURA	Work Office at Shonai River, Ministry of Construction
Ogawara Lake	1982	Geomorphological Land Classification Map of Ogawara Lake	M. OYA S. SUGIURA Y. HIRAI	Takase River Comprehensive Development Work Office, Ministry of Construction
Shonai Plain	1982	Geomorphological Land Classification Map of the Shonai Plain Illustrating Features of Flooding and Soil Liquefaction	M. OYA K. KOTODA K. WAKAMATSU S. KUBO	Work office at Sakata, Ministry of Construction
Abashiri River	1984	Geomorphological Survey Map of the Abashiri River Basin Showing Classification of Flood-Stricken Areas	M. OYA M. UMITSU S. HARUYAMA Y. HIRAI	Division of Development and Construction at Abashiri, Bureau of Hokkaido Development

Area	Year Published	Map Title	Author	Publishing Office
Padang Plain (Sumatra, Indonesia)	1983	Geomorphological Survey Map of Padang Area in West Sumatra Showing Classification of Flood-Stricken Areas	M. OYA	JICA
Nile River (Egypt)	1986	Geomorphological Land Classification Map of the Nile River Basin	M. OYA S. HARUYAMA	Waseda University
Chao Phraya River (Thailand)	1989	Geomorphological Survey Map of the Central Plain of Thailand Showing Classification of Flood-Inundated Areas	H. OKURA M. OYA S. HARUYAMA et al.	National Research Center for Disaster Prevention, Japan; Waseda University; Remote Sensing Division, National Research Council of Thailand
Kraseio River (Thailand)	1991	Geomorphological Survey Map of the Kraseio River Basin in the Western Part of the Central Plain of Thailand Showing Classification of Flood-Inundated Areas	H. OKURA S. HARUYAMA M. OYA et al.	National Research Center for Disaster Prevention, Japan; Waseda University; Remote Sensing Division, National Research Council of Thailand
Atlas	1993	Geomorphological Survey Maps Showing Classification of Flood-Stricken Areas	M. OYA	Waseda University

Jomon Period	Incipient: 12,000−10,000 B.P. Initial: 10,000−7,000 B.P. Early: Centered on the marine transgression of c.a. 6,000 B.P. Middle: 4,500−3,500 B.P. Late: 3500−3,000 B.P. Final: 3,000−2,300 B.P.
Yayoi Period	Early / Middle / Late: c.a. 350 BC to 300 AD
Kofun Period	Middle: 400−500 Late: 500 –700
Asuka Period	600−710
Nara Period	710−793 Capital: Nara
Heian Period	794−1191 Capital: Kyoto
Kamakura Period	1192−1336 Capital: Kyoto (Emperor), Kamakura (Shogun)
Azuchi-Momoyama Period	1336−1603 Capital: Kyoto
Edo Period	1603−1867 Capital: Kyoto (Emperor), Edo (Shogun)
Meiji Period	1868−1911 Capital: Tokyo
Taisho Period	1912−1925 Capital: Tokyo
Showa Period	1926−1988 Capital: Tokyo
Heisei Period	1989−Present Capital: Tokyo

Appended Table 2 Outline of Historic Chronology of Japan

Geologic Age		Years Before Present	Glacial Age			Sea Level (m)	Cultural Age of Japan
			Europe	Poland	Japan		
Quaternary	Holocene	0	Post-glacial			0	
		0.2				-3	Yayoi
		0.4					Jomon
		0.6				2 to 6	
		0.8					
		1.0				-40	
	Pleistocene	1.2	Würm	Vistulian	Hida II Hida I		
		1.4					
		1.6					
		1.8				-140	
		2.0					
		4				-100	
		6					
		8					
		10	Riss/Würm	Eemian		5 to 8	
		12					
		14	Riss			-100	
		16					
		18					
		20					
		22	Mindel/Riss				
		24					
		26	Mindel			-100	
		28					
		30					
		80	Günz/Mindel				
		90	Günz				

Appended Table 3 Glacial Age of the Quaternary

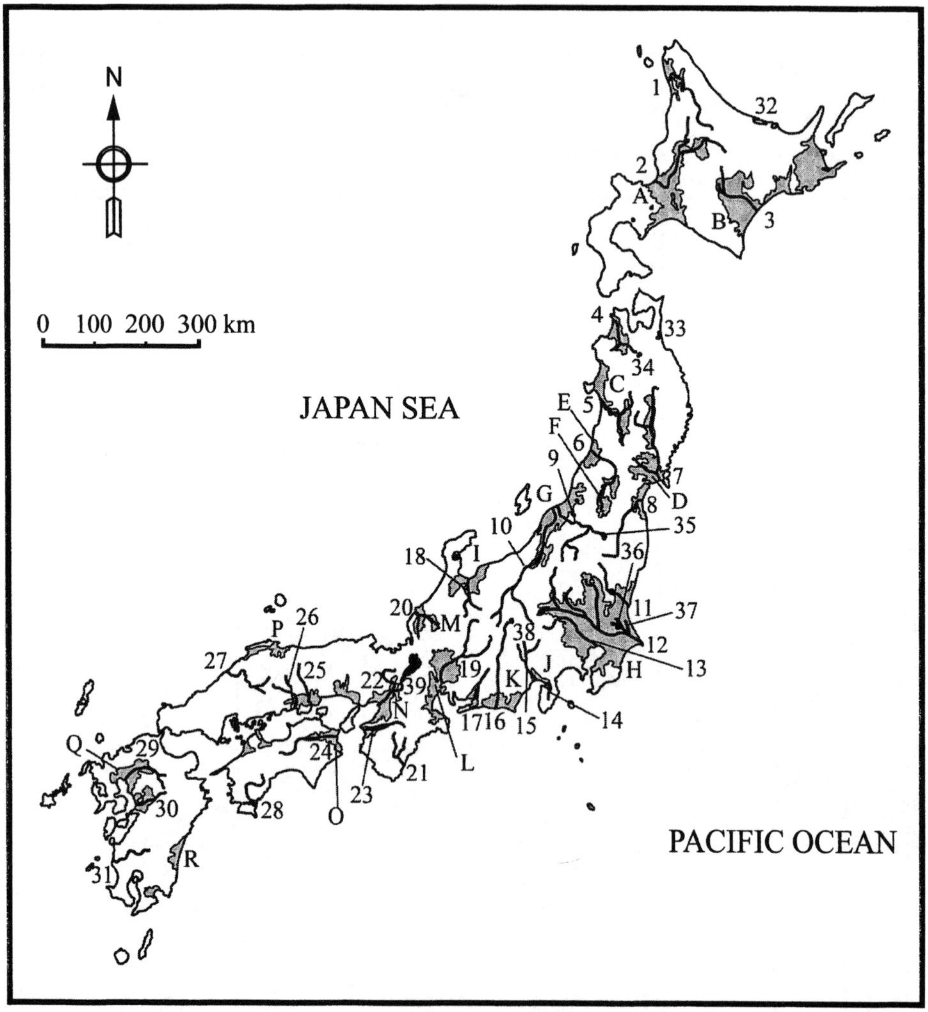

Appended Figure Rivers, lakes and plains in Japan

Rivers and Lakes:

(1) Teshio River, (2) Ishikari River, (3) Tokachi River, (4) Iwaki River, (5) Omono River, (6) Mogami River, (7) Kitakami River, (8) Abukuma River, (9) Agano River, (10) Shinano River, (11) Naka River, (12) Tone River, (13) Ara River, (14) Kano River, (15) Fuji River, (16) Tenryu River, (17) Toyo River, (18) Jintsu River, (19) Kiso River, (20) Kuzuryu River, (21) Kumano River, (22) Yodo River, (23) Kino River, (24) Yoshino River, (25) Asahi River, (26) Takahashi River, (27) Gouno River, (28) Shimanto River, (29) Chikugo River, (30) Shira River, (31) Sendai River, (32) Saroma Lake, (33) Ogawara Lake, (34) Towada Lake, (35) Inawashiro Lake, (36) Kasumigaura Lake, (37) Kitaura Lake, (38) Suwa Lake, (39) Biwa Lake.

Plains:

(A) Ishikari Plain, (B) Tokachi Plain, (C) Akita Plain, (D) Sendai Plain, (E) Shonai Plain, (F) Yamagata Intermontane Depression, (G) Niigata Plain, (H) Kanto Plain, (I) Toyama Plain, (J) Kano River Plain, (K) Mikatagahara Terrace, (L) Nobi Plain, (M) Kuzuryu Plain, (N) Osaka Plain, (O) Tokushima Plain (Yoshino River Plain), (P) Izumo Plain, (Q) Chikugo Plain, (R) Miyazaki Plain.

REFERENCES

AKAGIRI, Takekazu and WAKAMATSU, Kazue (1991): Damage of Soil Liquefaction Caused by Earthquakes and Use of Geomorphological Maps. Bulletin of the Geographical Survey Institute, Vol. 36. (J/E)

AKI, Koichi (1952): Suigai no Nippon (Flood damage in Japan). Iwanami-Sinsho. (J)

ASIAN RESEARCH SERVICE (1980): Asian Profile (Hong Kong), Vol. 8, No. 3. (E)

DEMEK, Jaromir (1972): Manual of Detailed Geomorphological Mapping. Academia. Publishing House of the Czechoslovak Academy of Sciences, C.S.S.R. (E)

DOUGAL, Merwin (1969): Flood Plain Management, Iowa's Experience. Iowa State University Press, U.S. (E)

EMBLETON, Clifford (1984): Geomorphology of Europe. The Macmillan Press Ltd. (E)

GEOGRAPHICAL SURVEY INSTITUTE (1947): Flooding in September, 1947.

Flooding Report for the Tone and Ara Rivers. Geographical Survey Institute. (E)

HIRAI, Yukihiro (1983): Lacustrine and Sub-lacustrine Microforms of Lake Ogawara with Relation to the Lake—Level Changes after the Maximum Post Glacial Transgression, Tohoku Chiri, Vol. 35. (J/E)

HIRAI, Yukihiro (1987): Structures and Origin of the Littoral Shelves at Lake Kasumigaura. Geographical Review of Japan, Vol. 60 (Ser. A), No. 12. (J/E)

HIRAI, Yukihiro (1989): Geomorphological Features and Development of Maritime Coastal Lakes in Japan. Geographical Review of Japan, Vol. 62 (Ser. A), No. 2. (J/E)

HIRAI, Yukihiro (1992): Environmental Changes at Coastal Lakes in Japan on the Basis of Geomorphological Land Classification and Land Use. Study Report of the Faculty of Education, Ehime University, Vol. 12, No. 2. (J/E)

HIRAI, Yukihiro (1994): Geomorphological Development of Lagoons in Japan. Study Report of the Faculty of Education, Ehime University, Vol. 14, No. 2. (J/E)

ICHINOSE, Yoshimi (1966): The Formation and the Outflow of the Debris in Landslip Areas —In the case of the Yoshino Valley, Part I. Miscellaneous Reports of the Research Institute for Natural Resources, Vol. 67. (J/E)

ICHINOSE, Yoshimi (1967): The Formation and the Outflow of the Debris in Landslip Areas —In the case of the Yoshino Valley, Part II. Miscellaneous Reports of the Research Institute for Natural Resources, Vol. 68. (J/E)

MCINTIRE, William and MORGAN, James (1959): Quaternary Geology of the Bengal Basin, East Pakistan and India. Bulletin of the Geological Society of America, Vol. 70. (E)

MOJSKI, Jozef (1980): Vistulian Stratigraphy in the Glaciated Area of the Polish Lowlands. Quaternary Studies in Poland, Vol. 2. (E)

NAKANO, Takamasa and OYA, Masahiko (1960): Relationship Between High Tide Caused by the Ise Bay Typhoon and Topography of the Plain. Geographical Survey Institute, Ministry of Construction. (J)

NAKANO, Takamasa, OYA, Masahiko and TAKEHISA, Yoshihiko (1962): Relation between Tsunami and Coastal Landforms in the Case of the Chilean Earthquake Tsunami, May 24th, 1960. Bulletin of the Geographical Survey Institute, Vol. VII, Parts 24. (E)

NATIONAL ASTRONOMICAL OBSERVATORY (1980): Chronological Science Tables. (J)

NATIONAL RESEARCH CENTER FOR DISASTER PREVENTION (1969): Quaternary Tectonic Map of Japan. (J/E)

OYA, Masahiko (1956): Topographical Survey Map of the Kiso River Basin Showing Classification of Flood-Stricken Areas. Appended map of the Resources Council Reference Data, Prime Minister's Office. (J)

OYA, Masahiko (1960): Topographical Survey Map of the Middle and Lower Courses of the Kano River Basin Showing Classification of Flood-Stricken Areas. Geographical Review of Japan, Vol. 33, No. 3. (J/E)

OYA, Masahiko (1967): Geomorphological Study of Flooding Immediately Downstream from Pa Mong in the Mekong River Basin. The Committee for Co-ordination of Investigations of the Lower Mekong Basin, ECAFE (U.N.). (E)

OYA, Masahiko and NAKAMURA, Shukue (1969): Geographical Study on the Inundation Caused by Local Rainfall in the Neya River near Osaka. Miscellaneous Reports of the Research Institute for Natural Resources, Vol. 72. (J/E)

OYA, Masahiko (1970): Land Use Control and Settlement Plans in the Flooded Area of the City of Nagoya and Vicinity, Japan. Geoforum (Germany), 1970, No. 4. (E)

OYA, Masahiko and MOCK, F. J. (1970): Geomorphology of the Naktong River Basin. FAO, U.N. (E)

OYA, Masahiko (1971): Geomorphological Flood Analysis of the Naktong River Basin, South Korea. Report to FAO, U.N. (E)

OYA, Masahiko (1972a): Recent Trends of Landform Classification Studies in Japan. Geographical Sciences, Vol. 17. (E)

OYA, Masahiko (1972b): Contribution of Physical Geography on Flood Control. Miscellaneous Problems on Geography. Hiroshima University. (J/E)

OYA, Masahiko (1973a): Relationship between Geomorphology of the Alluvial Plain and Inundation. Asian Profile (Hong Kong), Vol. 1, No.3. (E)

OYA, Masahiko (1973b): Basic Form of the Combination of Geomorphologic Elements in the Alluvial Plain. Gakujutsu Kenkyu (Academic Studies), The school of Education, Waseda University, Vol. 22. (J/E)

OYA, Masahiko (1977a): Comparative study of the Fluvial Plain Based on Geomorphological Land Classification. Geographical Review of Japan, Vol. 50 (Ser. A), No. 1. (E/J)

OYA, Masahiko (1977b): Applied Geomorphological Study on the Selection of a Bridge Site along the Jamuna River in Bangladesh. National Geographer (India), Vol. 12 No.2. (E)

OYA, Masahiko (1980): Multifunctional Uses of Geomorphological Survey Maps in Identifying Flood-Stricken Areas and Land Types. National Geographer (India), Vol. 15, No. 2. (E)

OYA, Masahiko (1981):Physical Geographic Features of the River in Hokkaido. Gakujutsu Kenkyu (Academic Studies), The school of Education, Waseda University, Vol. 30. (J)

OYA, Masahiko and HARUYAMA, Shigeko (1987a): Flooding and Urbanization in the Lowland of Tokyo and Vicinity. Natural Disaster Science, Vol. 9, No. 2. (E)

OYA, Masahiko (1987b): Distinctive Features of the Rivers and Fluvial Plains in the Semi-Fridge Zone—with Particular Reference to the Vistula River in Poland. Gakujutsu Kenkyu (Academic Studies), The school of Education, Waseda University, Vol. 36. (J)

OYA, M. et al. (1988a): Studies on the Geomorphological Features of the Fluvial Plains in Japan, Focusing on Distribution, Geomorphological Land Classification and Its Application. Geographical Review of Japan, Vol. 61 (Ser. B), No. 1. (E/J)

OYA, Masahiko (1988b): Plan Geomorphologique de Padang et de ses Environs, Sumatra Ouest (Indonesie), Montrant les Regions Susceptibles d'etre par des Inondations. Travaux et Documents de Geographie Tropicale, GEGET-CNRS, Bordeaux, Vol. 61, 1988. (F/E)

OYA, M. et al. (1989a): Geomorphological Land Classification Map of Lake Kasumigaura and its Vicinity, East of Tokyo, Geografia Fisica e Dinamica Quaternaria, Vol. 12, No. 1 (Italy). (E/I)

OYA, Masahiko (1989b): Report of the Working Group of the Association of Japanese Geographers on "Geomorphology of Rivers and Coastal Plains." Geographical Review of Japan, Vol. 62 (Ser. A), No. 2. (J)

OYA, Masahiko (1990): Report on the activities of the Working Group on the Geomorphology of the River and Coastal Plains, International Geographical Union. Gakujutsu Kenkyu (Academic Studies), The School of Education, Waseda University, Vol. 39. (E)

OYA, Masahiko (1993): Atlas, Geomorphological Survey Maps Showing Classification of Flood-Stricken Areas. Waseda University Press. (J/E)

OYA, M. et al. (1993): International Symposium on Flood Disaster Reduction in Southeast Asia (Nov. 1991). Geojournal Vol. 31, No. 4. (E)

OYA, Masahiko and KUBO, Sumiko (1993): Geomorphological Survey Map of the Yodo River Showing Classification of Flood-Stricken Areas. Yodo River Work Office, Ministry of Construction. (J/E)

OYA, M. et al. (1994): Applied Geography for the Mitigation of Natural Hazards and Conservation of the Environment. (Kokon Shoin). (J)

OYA, Masahiko (1995): Geomorphological Maps of Alluvial Plains and their Utilization for Mitigation of Natural Hazards: Flooding and Soil Liquefaction. Geographical Review of Japan, Vol. 68 (Ser. B), No. 2. (E)

OYA, Masahiko (1996): Brief report of the International Congress on Geomorphological Hazards in the Asia-Pacific Region (Sept. 1993). Geojournal Vol. 38, No. 3. (E)

OYA, Masahiko and SAKA, Yukiyasu (1997): Yoroppa no Chikei. Translated from Geomorphology of Europe edited by EMBLETON, Clifford in 1984, (Taimeido). (J)

TADA, F., OYA, M. et al. (1956): Reconnaissance Topographical Survey on Areas Subject to Flooding, Part 1. Resources Council Reference Data, Vol. 46, Resources Council, Prime Minister's Office. (J)

TADA, F., OYA, M. et al. (1957): Reconnaissance Topographical Survey on Areas Subject to Flooding, Part 2—The Relationship Between Topography and Flood Type in the Chikugo River. Material of the Resources Bureau Science and Technology Agency, Vol. 13, Resources Bureau, Science & Technology Agency. (J)

TADA, Fumio and OYA, Masahiko (1959): Flood Types and Classification of Topography. Proceedings of the IGU Regional Conference in Japan 1957. (E)

TADA, Fumio, OYA, Masahiko and ICHINOSE, Yoshimi (1959): Research on the Isahaya Suigai (Flooding). Reconnaissance Topographical Survey on Areas Subject to Flooding, Part 3. Resources Bureau, Science and Technology Agency. (J/E)

UCHIDA, Kazuko (1976): Applying the Standard of the International Land Classification of IGU to Japan. Map, Vol. 14, No. 3. (J/E)

WALTER DE GRUYTER (1980): Allgemeine Klimageographie. (G)

WAKAMATSU, Kazue (1980): Geomorphological Consideration of Site Conditions in Soil Liquefaction. Asian Profile (Hong Kong), Vol. 8, No. 3. (E)

WAKAMATSU, Kazue (1991): Maps of Historic Liquefaction Sites in Japan. Tokai University Press. (J/E)

INDEX

A

Abashiri 69, **71**

Aerial photograph **8**, 103, 137

Air Dingin River 131, 132, **135**

Akaishi Mountains **37**

Alluvial Plain 1, 5, 8, 12, **41**, 43, 60, 103, 105, 111, 124, 128, 140, 142, 143, 144

Applied geographical studies **12**

Arau River **131**, 132

Ariake Sea **25**, 28, 45

Aricha **141, 143**

Artificially filled field **9**

Asama Volcano **97**

Asan Mountains **82, 85**

B

Back-marsh 1, **6**, 8, **9**, 17, 23, **41**, 42, **43**, 49, 55, 57, 69, 71, 73, 103, 105, 111, 116, 124, **129**, 137

Bahadurabad 141, 143

Bai-u front **91**

Bangladesh 131, 135, 136, 140, 141

Brahmaputra-Jamuna River 135, 136, 140, **142, 143, 144**

C

Central Japan Alps **37**

Central Plain of Thailand 6, **131**

Chikugo River 6, **45**, 47

Chilean Earthquake **29**, 31

Costal lake **101**

Crustal movement 14, 17, **37**, **47**, 52, 79, 88, 101, 134, 142, 143

D

Delta 6, 8, **9**, **17**, 23, 25, 34, **41**, **42**, **43**, 45, 49, 55, 69, 71, 72, 73, 86, **100**, 101, 103, 105, 108, 111, 112, 124, 126, **129**, 132, 137, 140, 141, 142, 143

Dhaka **140**

Dutch **131**

E

ECAFE **6, 131**
(U.N. Economic Commission of Asia and Far East)

Edo Period **1**, 42, 81, **98**, 111, 156

ESCAP **6**

Eustatic movement **42**, 106

F

Fan **9, 41, 129**

Flood type **6**, **37**, 47, 52, 60, 100

Floodwater **6**, 28, 47, 49, 55, 57, 60, 69, 82, 100, 112, 116

Former river course **9**

Fossa Magna **79**

Fuji Volcano 49, **79**, 91

Furu-Tone River 1, **111**

G

Gabargaon 141, **143**

Ganges Plain **142**

Ganges River **142**

Geographical Survey Institute **5**, **8**, 86, 103

Geomorphological Land Classification Map **8**, 12, 60, 109, 121, 124, 131, **136**

Geomorphological Survey and Mapping **11**

Geomorphological Survey Map **1**, 5, **6**, 8, 17, 55, 61, 63, 114, 131, 132

Geomorphology of River and Coastal Plains Working Group **12**

H

Han River 45, **57**, **60**

Hida Mountain **37**, 42

Himalayan Orogenic Movement **142**

Hokkaido 31, **67**, 72, 79, 101

Holocene 63, 73, **101**, 105, 116, 141, 142

Hooghly River 140, **142**

Hwang River **57**

Hyogo Prefecture Nanbu Earthquake **126**
(Great Hanshin Earthquake)

I

Ice jam **72**, **76**

IGU (International Geographical Union) 5, **11**, 12

Indonesia 6, 28, **131**

Inner Zone **79**, 85, 86, 88, 90

International Geomorphological Map of Europe **11**, **13**

Inuyama **43**

Isahaya Suigai **98**

Ise Bay Typhoon **17**, 23, 25, 28, 43, 97, 117

Ishikari Plain **67**, 69

Ishikari River **67**, 69

J

Jamuna River **143**

Jamuna River Bridge Construction Project Team **136**, 141

Japanese archipelago 1, 28, 29, **37**, **39**, **79**

JICA 6, **131**
(Japan International Cooperation Agency)

Jomon Transgression **42**, 105, 106, 108, **156**

K

Kano River **49**

Kano River Typhoon **49**, 111, **112**, 114

Kanto Earthquake **37**

Kanto Plain 1, **37**, 103, 111

Kasumigaura Lake 101, **103**, **105**, 106, 109

Katsushika Ward **111**

Kawachi Plain **114**

Kiso Mountains 42, **94**, 97

Kiso River 5, 17, 23, **42**, 43, 45, 47, 71, 94

Kitaura Lake 101, **103**, 105, 106

Knick point 132, **134**

Korean Peninsula **52**, 55

Krakatou Volcano **29**

Kuranji River **131**, **132**, 134, 135

L

Lacustrine Terrace **101**, 105, 106, 108, 109

Land collapse 52, 72, **91**, **94**, **97**, 98

Laos **61**, 131

Landslide **43**, 45, **71**, 91, **94**, **97**, **98**, 100

Littoral Shelf **101**, 105, 108, 109

M

Maduhupur Jungle Terrace **140**, 141, 143

Median Dislocation Line **79**, 81, 88
Meghna River 137, **141**
Mekong River 6, **61**, 63
Micro-topography 5, **6**, 12, 121
Ministry of Construction **6**, 25, 28, 86, **117**
Mogami River 47, 77, **124**, **126**
Monsoon **39**, 91
Moor **69**

N

Nagoya 17, 20, 23, 25, **117**
Naka River **111**, **112**, 114
Nakdong River 45, **55**, 57
National Research Center for Disaster Prevention **6**
Natural levee 6, 8, **9**, 14, 17, 23, **41**, 42, **43**, 45, 49, **55**, 57, 60, 63, **69**, **71**, 72, 73, 86, 90, 103, 105, 108, 111, 114, 116, 124, **129**, 132, 137, 141, 143
Neya River 114, **116**, 117
Niigata Earthquake 121, **126**, 128
Nobi Plain **17**, 25, **42**, 45, **71**, 124
Northern Japan Alps **37**

O

Ogawara Lake 101, **106**, **108**
Old Brahmaputra River **140**, **143**
Osaka 17, 39, 42, **114**, 126
Outer Zone **79**, 81, 82, 85, 86, 88

P

Pa Mong Dam **61**
Padang 6, **131**, **132**, **134**, 135
Peat-bog **69**, 72
Peri-glacial region **72**

Q

Quaternary Period **37**, 91, **94**

R

Reclaimed land **9**, **129**
Reconnaissance Topographical Survey Map **5**, **17**
Relief energy 43, 45, **94**
Rias-type **31**

S

Sand spit **9**, **129**
Sand dune **9**, **129**
Sanriku Coast **28**, 29, 31
Shikoku Mountains **81**, 82, 85
Shokotsu River 69, **71**
Shonai Plain 122, **124**, 126, 128
Sirajganj 136, 141, **143**, **144**
Soil liquefaction 12, **121**, 122, **124**, **126**, 128
Southern Japan Alps **37**
Sundarbans **141**
Sylhet Depression **141**

T

Taradake Volcano **98**, 100
Thailand **61**, 131
Tidal wave 6, 9, **20**, **23**, **25**, 28, 39, 109
Tippera Surface **141**, 142
Tista River **140**, 143
Tokoro River 69, **71**
Tokyo 1, 12, 17, 37, 39, 42, 49, 103, **111**, 112
Tone River 1, 39, 98, 103, 105, 106, 109, **111**, 114
Topographical Survey Map of the Lower Part of the Kiso River Basin **5**, **17**
Toyo River 79, **88**, 90

Tsunami 6, **28**, 29, **31**, 34, 39, 86, 109
Typhoon **1**, 6, 17, 20, 23, 25, 28, 31, 39, 43, 49, 67, 72, 81, 82, 90, 91, 97, 109, 111, 112, 114, 117
Typhoon Catherine **1, 111**

U

UNESCO **12**

V

Valley plain 6, **9**, 34, 100, 103, 105, **108**, 124, **129**, 137, 143
Vientiane Plain **63**
Vistula River **73**

W

Western Barind Terrace **140**
Würm Ice Age 71, **106**, 134, **141**

Y

Yamato River 67, **114, 116**
Yodo River **114**, 116, 126
Yoshino River 79, **81, 82**, 85, 86, 90
Yubetsu River 69, **71**

AUTHOR'S PERSONAL HISTORY

Masahiko Oya

Born May 1, 1923, in Nagoya, Japan.
Studied Human Geography at Nagoya University and received a Bachelor of Arts Degree in 1951.
Completed a post-graduate course in physical geography at the University of Tokyo in 1958.
Worked for the Geographical Survey Institute, Ministry of Construction.
Received a Ph.D. in science from Tokyo Metropolitan University in 1976.
Was an associate professor at both Aichi Prefectural University and Waseda University, and a professor at Waseda University until 1994.
During this period, worked for international organizations such as ECAFE (ESCAP) U.N., FAO U.N. and ITC (International Institutes for Aerial Survey and Earth Science) in the Netherlands.
Currently professor emeritus at Waseda University, honorary director of Katsushika City Museum and technical advisor for Nikken Consultants Inc.

COLOR PLATE SECTION

Photo 1 Coastal embankment destroyed by a tidal wave caused by the Ise Bay Typhoon (1959)

Photo 2 Floodwater covers paddies and nearly reaches a farmer's house on a natural levee near the mouth of the Kiso River. (1959)

Photo 3 This railway for transporting lumber was destroyed by land collapse along the Otaki River, a branch of the Kiso River. (1968)

Photo 4 Mudflow in the upper reaches of the Kiso River (1968)

Photo 5 Professor Fumio Tada examines a climatic terrace along the Nakdong River, South Korea. (1970)

Photo 6 A natural levee along the Nakdong River near Seoul (1970)

Photo 7 Deltaic area of the Nakdong River (1985)

Photo 8 A Thai engineer measures new deposits on a natural levee following the flooding of the Mekong River. (1966)

Photo 9 A Laotian points to a mudstain on a post to demonstrate floodwater levels in a village near Vientiane. (1966)

Photo 10 Lowest reach of the Chikugo River during low tide (1970)

Photo 11 Aso Volcano in Kyushu

Photo 12 Soil liquefaction created by the Great Hanshin Earthquake caused the Yodo River embankments to incline toward the river. (1995)

Photo 13 The author poses by the Shokotsu River in Hokkaido. (1983)

Photo 14 An erosional plain near the Vistula River in Poland (1974)

Photo 15 Wotswabeck Dam on the Vistula River, Poland (1981)

Photo 16 The natural levee of a former river course in the Ganges Plain of Bangladesh is covered by trees and houses. Back swamps are used as paddy fields. (1974)

Photo 17 Sundarbans Forest, East Pakistan (now Bangladesh) (1962)

Photo 18 Bank erosion along the Brahmaputra-Jamuna River in Bangladesh (1975)